FROM THE PUBLISHER OF
FineScale MODELER MAGAZINE

FAMOUS SPACESHIPS OF FACT AND FANTASY

2ND EDITION

KALMBACH BOOKS

© 1996 by Kalmbach Publishing Co. All rights reserved. This book may not be reproduced in part or in whole without written permission of the publisher, except in the case of brief quotations used in reviews. Published by Kalmbach Publishing Co., 21027 Crossroads Circle, Waukesha, WI 53187. Telephone: (414) 796-8776.

Printed in the United States of America

Book design: Kristi Ludwig

Publisher's Cataloging in Publication
(Prepared by Quality Books Inc.)

Famous spaceships of fact and fantasy / [edited by Terry
  Spohn] ; from the publishers of FineScale Modeler
  magazine. — 2nd ed.
    p. cm.
  Includes bibliographical references and index.
  ISBN 0-89024-563-0

  1. Space vehicles—Models. I. Spohn, Terry. II. Title:
FineScale modeler.

TL844.F35 1996   629.47'022'8
                            QBI95-20694

CONTENTS

Apollo, by James E. Oberg 5

"The *Eagle* Has Landed!" by David Senechal 13

Saturn V, by James E. Oberg 19

Beginner's Guide to Plastic Kit Modeling, by Harold A. Edmonson 27

Build Apollo by Subassembly, by Bob Hayden 34

The Ultimate Starship, by Richard G. Van Treuren 37

Kit Conversion: *Enterprise* to Tug, by Don Klein 39

Decaling—A Most Rewarding Enterprise, by Don Klein 41

Lighting AMT's Klingon Battle Cruiser, by Mark P. Wilson 43

Painting and Mounting a Klingon Bird of Prey, by Rusty White 47

Wiring Ertl's "Next Generation" USS *Enterprise*, by Shawn Marshall 51

Accurizing Ertl/AMT's Starship *Enterprise*, by Chris Paveglio 58

Making *Galileo* Accurate, by Marc Millis 63

Assignment Sound Stage, by Richard G. Van Treuren 67

Weathering is Key to X-wing, by Don Klein 71

Masking and Painting the TIE, by Don Klein 76

Viper and Raider, by Chris Tietz 80

Fiber Optics Light the Raider, by Bob Hayden 85

Battle Damage Adds Realism to Viper, by Bob Hayden 88

Science-Fiction Modeling with Junk, by Brian Tremblay 91

The Ships of *Star Wars*, by Chris Tietz and George Elrick 93

Index 103

Although the entire Apollo program drew heavily upon technology developed by the U.S. armed forces, NASA and the astronauts themselves emphasized the pacific and imaginative aspects of lunar exploration. Neil Armstrong's first words as he took man's first step on the moon were "That's one small step for a man; one giant leap for mankind." A plaque on the LM ladder read "Here men from the planet Earth first set foot upon the Moon, July 1969 A.D. We came in peace for all mankind."

APOLLO

"One giant leap for mankind"

JAMES E. OBERG

The Apollo spaceships opened the grandest frontier in the history of exploration, and they blazed the farthest and longest trails ever ridden by explorers. The quarter-million-mile voyage to the moon will doubtlessly someday be eclipsed by million-mile hops to the planets and nearby stars, but Apollo will always remain symbolic of the first visit by men to another world beyond the Earth. Project Apollo proved that humanity is no longer bound to its planet of birth.

Apollo spaceships also played an instrumental role in the triumphant triple expeditions to the space station Skylab in 1973–1974 that set space endurance records

Rendezvous and docking maneuvers are among the most difficult tasks of space flight. Computers handle the mathematical calculations, but rendezvousing pilots must carefully align the spacecraft during the docking. Apollo 9 gave NASA practice in docking the CSM and LM while in Earth orbit. Here are the Apollo 9 LM and CSM as seen from one another.

and unveiled space secrets. More space history was made a year later during the epochal Soviet-American space linkup between a Russian Soyuz spaceship and the last American Apollo spacecraft.

The lunar expeditions of 1969–1972 actually made use of two separate manned space vehicles, both included in the project named Apollo. No more unlikely set of space twins could be imagined—the conical, sleek Command Module and the awkward, misshapen two-staged four-legged Lunar Module. The Command Module (or CM) remained attached to the cylindrical Service Module (or SM) which provided the astronauts with power, air, and propulsion. The Lunar Module (or LM) made the actual moon landing while the Command and Service modules remained in lunar orbit many miles overhead.

The Command Module was the section that ultimately returned to earth. During the mission, it provided living and working space for the three-man crew. The astronauts lay side by side on couches, with the main control panel stretched out above them. All the dials, switches, displays, buttons, and other gauges covered a main console 9 feet wide by 3 feet high.

The leftmost seat was occupied by the flight commander, the senior astronaut. The middle seat was for the Command Module pilot, whose duties involved guidance, navigation, and special Command Module functions. The right seat was for the Lunar Module pilot, who was also responsible for monitoring engineering systems on the Command Module (on flights without a Lunar Module, the man in this seat was a specialist in engineering or scientific experiments). The middle seat was removable in flight to provide more room.

The main entry hatch was positioned in the wall of the Command Module so the center astronaut could reach over his head and unlatch it quickly. At the feet of the three crewmen was a "lower equipment bay" containing lockers and navigational sighting equipment. From this section, a small tunnel ran up behind the control panel to the nose of the Command Module.

During moon flights, this tunnel led to the Lunar Module; on Skylab flights, it led to the space station itself; and on Apollo-Soyuz, it led to the docking module and thence to the Soviet spacecraft. The Command Module pilot was responsible for opening and closing the airlock hatches at the nose of the Command Module, and for handling the removable docking equipment (the probe and socket).

Aft of the Command Module was the Service Module, and these two units were so closely related that they were generally given an abbreviation of their own—CSM—or Command/Service Module. The Service Module provided all the power to the Command Module, via fuel cells. It contained the Spacecraft Propulsion System, or SPS, which allowed the astronauts to steer through space. Indeed, without the SPS, the astronauts would have been unable to come home. Most of the volume in the Service Module was taken up by fuel tanks for the rocket engines.

The Service Module was only jettisoned at the last moments of flight, when batteries in the Command Module could carry the electrical load for an hour or so.

For the lunar landing flights, a third module was employed: the Lunar Module. It was the craft that

Parting from the CSM was always exciting for the astronauts in the LM. Neil Armstrong and Edwin Aldrin left the Apollo 11 CSM pointing straight after them high above the desolate Sea of Fertility. David Scott and James Irwin enjoyed a side view of the Apollo 15 CSM (with the Scientific Instrument Module bay exposed) shortly before the lunar descent.

made the landing, while the earth-return vehicle remained—for fuel efficiency—in orbit above the moon under control of the Command Module pilot. Two astronauts stood in front of a control panel in the Lunar Module. There were no chairs or couches since the gentle accelerations in the weak lunar gravity field did not require them. Each astronaut had a window to look out and downward from, which would have been awkward from any other posture but standing. The main control panel was between them, although both walls and the ceiling were also covered with instruments and switches. Below the central panel was the moonwalk hatch, located so that an astronaut with a life support backpack could crawl backwards through the door (which opened inwards) and onto a porch in front of the module. From the porch, a ladder led to the ground.

Most of the equipment for the activity on the lunar surface was located in lockers on the outside of the Lunar Module. The astronauts could remove this material once they were on the surface. On later flights, a folding moon jeep was included: The crew stood at a safe distance, pulled a lanyard, and watched their small four-wheeled "lunar rover" unfold itself from the side of their landing craft.

A commonly asked question involves how the astronauts got their heavy moon rock boxes and bags of film canisters back into the spaceship. They did not carry them up the ladder one-handed, but instead used what they called a "clothesline" pulley arrangement. One astronaut in the Lunar Module pulled on a line while the man on the surface attached the boxes to hooks on the line, which led from where he was standing up through the open hatch.

A second question is how they slept. They did not stand up all the time, of course. For sleeping, they strung two hammocks: One ran from right to left across the cabin, and the other ran from front to back. Since the sun did not set on the moon, the astronauts had special shutters for their windows.

The Lunar Module consisted of two stages. Most of the fuel tanks, the descent engine, the landing legs, and the lunar surface equipment were all left behind when the astronauts left the moon in the "ascent stage." This was, so to speak, hardly more than a flying telephone booth. The rocket engine was under a cover right behind the crew, and the fuel tanks were strapped to either side of the cabin.

Much of the exterior of the LM was covered with protective Mylar foil. Various color patterns were seen on the different missions, with orange, silver, and yellow foil contrasting with surfaces painted in black and white. The colors reflected or absorbed the sun's heat differently, and a good choice of color scheme (combined with the appropriate orientation toward the sun) helped take the load off the spacecraft's cooling system. For example, on early lunar landings, the astronauts touched down soon after lunar sunrise at the landing site, when shadows were the longest (this facilitated the pilot's job of gauging the touchdown on unknown terrain). Later missions put down nearer to the lunar noon, which was an excellent time for orbital photographic reconnaissance but put higher loads on the LM's air conditioning.

Moon mission. Now let's take a look at a typical moon mission, and see how the components worked together. At launch, the Command Module was in place at the top of the Saturn rocket. Only an escape tower extended above the CM. Had there been an emergency or aborted take-off, a solid-fuel rocket would have pulled the spacecraft up and away from the Saturn. In the entire history of the Apollo program, the escape system was never needed.

The Lunar Module was positioned below the Command and Service modules inside a "garage."

The Apollo and the third stage of the Saturn were placed into a low circular orbit of the earth, where additional navigational checks were conducted. Using the "parking orbit" technique also gave flight planners great flexibility in the event of a launch delay. After one or two orbits, the last stage of the Saturn was fired again to increase speed from 17,000 mph to 25,000 mph, sending Apollo to the moon.

The next order of business was to link up with the Lunar Module. The Command/Service Module was slowly flown off the LM "garage," causing the springloaded garage walls (SLA panels) to fly free. The Command Module pilot fired his thrusters to turn the spacecraft around. The exposed Lunar Module became the docking target, and the nose of the Command Module was piloted to attach to the roof of the LM. The connection secure, an order was sent to the LM tie-down latches to release the now-complete three-module spaceship.

The Saturn rocket's third stage and the walls of the LM garage drifted along in orbit. Ultimately the third stage was directed to hit the moon, providing known calibration impacts for lunar seismographs. During the three-day coast up and out to the moon, the astronauts checked out the Lunar Module and made midcourse corrections as needed.

As the 50-ton spacecraft (most of the weight was fuel) whipped around the back side of the moon, on what would have been a giant figure-8 orbit in space heading back toward earth, the astronauts fired their main rocket engine to slow down their speed. This allowed the weak lunar gravity to hold them back in an orbit circling the moon. After another day or two of navigational checks, two of the three astronauts separated the LM from the CMS (where one astronaut remained, ready if need be to return alone to earth).

Apollo 10 CSM was lifted for mating to Lunar Module Adapter at KSC's Manned Spacecraft Operations Building. Ablative heat shield (protected during shipment by blue coating) covered the CM; reflective material covered cone for flight.

The LM's descent engine was used to cut the orbital speed enough to permit the moon's gravity to pull the lander closer. Approaching the landing site at low altitude, the main descent engine continued to fire at a rate precisely calculated so that the lander's velocity reached zero at the exact moment its altitude reached zero. Any leeway either way could have spelled disaster. If the astronauts had detected a bad guidance system, however, they could have hit the "abort" button, lighting the ascent engine and dropping the entire landing stage to its destruction while attempting themselves to get back into a stable lunar orbit where the Command/Service Module could recover them. Luckily, this failure never happened.

Aided by radars and by their computer, and using their own eyesight out the windows, the astronauts guided the LM to a final landing. Probes extending down from the landing legs alerted the crew to imminent contact, although in practice the astronauts ascertained this information by watching their spacecraft's shadow out the window.

Activities on the moon were two-fold: collect samples and implant instrument packages. On the early expeditions, the two astronauts went on foot; later they used the electric moon jeep, which extended their range to tens of miles. Hundreds of pounds of carefully selected and documented samples were collected. At five sites, the nuclear-powered science stations called ALSEP (Apollo Lunar Science Experiment Package) were set up, with seismometers, magnetometers, ion detectors, and other instruments. These stations sent back coordinated readings for years, and were finally turned off in 1977 when their isotope power units began to run low.

The launch from the moon was one of the tensest moments of the missions, since there were no backup units and no chance of rescue. The design of the Lunar Module's ascent stage proved to be highly reliable, and all six lunar ascents were flawless. Hours after blasting off from the moon, the two moon walkers linked up in lunar orbit with the Command Module and transferred themselves and their precious film and samples. The LM itself would be detached and sent crashing to the moon's surface.

The orbiting CM had not been idle, either. From orbits of 180 to only 20 miles (depending on the mission phase), batteries of high-resolution cameras surveyed the moon, while on later flights a laser altimeter measured the shape of the surface. Small sub-satellites were jettisoned to continue lunar studies.

Leaving circumlunar orbit was the next critical procedure, as the Command/Service Module's main engine had to perform flawlessly. It fired while the astronauts were behind the moon, flinging them up and out of the moon's gravity field to where the earth's gravity would drag them back. For three days the Apollo fell, picking up speed from a few thousand miles an hour to 25,000 mph right before atmospheric entry.

A scheduled two-hour spacewalk by Russell Schweickart during Apollo 9 was canceled after he suffered nausea and vomiting, but he was able to test the Extravehicular Activity suit under space conditions from the open CM hatch.

Before entry the faithful Service Module was cast off, exposing the heat shield at the base of the Command Module. The capsule decelerated and heat built up during the dramatic final moments of flight. All went well. The CM slowed to only a few hundred miles per hour, slow enough for its parachute to release for the final descent to an ocean splashdown, where recovery ships were waiting.

This summarizes the standard flight profile. Of course, things didn't always work out to the standard. Some things went wrong. Miscalculations were made. Disasters happened, or were barely averted.

The worst Apollo disaster took place not in flight but in training. On January 27, 1967, Virgil I. Grissom, Edward H. White, and Roger B. Chaffee, the first Apollo crew, were killed during a ground test at Cape Canaveral. Fire broke out in the capsule, and the noxious fumes from the flames asphyxiated the astronauts.

Major changes were made in the Apollo Command Module, especially in regard to the fire danger of a pure oxygen atmosphere. The pure oxygen had been selected over a two-gas system using oxygen and nitrogen because it was simpler, lighter, and easier. It greatly facilitated space walks, making "pre-breathing" to clear the tissues of nitrogen unnecessary. But it heightened the danger of fire, and a series of miscalculations had killed three men.

The first change involved the use of a mixed atmosphere for all ground tests and for launch as well. As the capsule climbed into space—where fires in zero gravity were less likely and would be less disastrous—the air gradually was changed to a pure oxygen composition. Also, a quick-opening hatch—which would have saved the Apollo 1 crewmen—was added. Materials to be used inside the capsule underwent rigorous fire tests.

Another major disaster almost occurred after an oxygen tank in the Service Module exploded when Apollo 13 was halfway to the moon. The landing was canceled, and for two tense days even the lives of the crewmen were in doubt. All power in the Command/Service Module was crippled, so the astronauts had to live in the Lunar Module and use

MISSION CHRONOLOGY, APOLLO/SATURN

Mission number	Dates	Duration	Earth orbits	Moon orbits	Crew-members	Capsule, module names, serials	Booster serial, mission description
Apollo 4	Nov. 9, 1967	8 hrs 37 min	3	—	unmanned	CM017	Saturn 501. Engine tests, first flight test of the Saturn V
Apollo 5	Jan 22, 1968	7 hrs 50 min	5	—	unmanned	LM1	Saturn SA-204. Test Lunar Module and Lunar Module propulsion systems in earth orbit
Apollo 6	Apr 4, 1968	9 hrs 56 min	3	—	unmanned	CM 020	Saturn 502. Second test of the Saturn V launch vehicle
Apollo 7	Oct. 11 to 22, 1968	260 hrs 9 min 3 sec	163	—	Schirra Eisele Cunningham	CM 101	Saturn SA-205. Docking exercises, safety and reliability tests of Command Module systems
Apollo 8	Dec. 21 to 27, 1968	147 hrs 0 min 42 sec	1.5	10	Borman Lovell Anders	CM103	Saturn 503. Lunar orbital flight first manned Saturn V
Apollo 9	Mar 3 to 13, 1969	241 hrs 0 min 54 sec	151	—	McDivitt Scott Schweickart	CM 104 *Gumdrop* LM 3 *Spider*	Saturn 504. Test Command Module and Lunar Module in earth orbit
Apollo 10	May 18 to 26, 1969	192 hrs 3 min 23 sec	1.5	31	Stafford Young Cernan	CM 106 *Charlie Brown* LM 4 *Snoopy*	Saturn 505. Lunar orbit, examine landing sites from altitude of 50,000 feet
Apollo 11	Jul 16 to 24, 1969	195 hrs 18 min 35 sec	1.5	30	Armstrong Collins Aldrin	CM 107 *Columbia* LM 5 *Eagle*	Saturn 506. Lunar landing experiments, soil samples, deployment of EASEP package
Apollo 12	Nov 14 to 24, 1969	244 hrs 36 min 25 sec	1.5	45	Conrad Gordon Bean	CM 108 Yankee *Clipper* LM 6 *Intrepid*	Saturn 507. Landing at site of Surveyor 3, exploration of Ocean of Storms, deployment of ALSEP package
Apollo 13	Apr 11 to 17, 1979	142 hrs 54 min 41 sec	1.5	—	Lovell Swigert Haise	CM 109 *Odyssey* LM 7 *Aquarius*	Saturn 508. Explore Fra Mauro crater. "Houston, we've had a problem here." Mission aborted due to ruptured SM oxygen tank.
Apollo 14	Jan 31 to Feb 9, 1971	215 hrs 1 min 57 sec	1.5	34	Shepard Roosa Mitchell	CM 110 *Kitty Hawk* LM 8 *Antares*	Saturn 509. Explore Fra Mauro crater
Apollo 15	Jul 26 to Aug 7, 1971	295 hrs 11 min 53 sec	1.5	74	Scott Worden Irwin	CM 112 *Endeavor* LM 10 *Falcon*	Saturn 510. Explore Sea of Rains, deploy lunar sub-satellite from lunar orbit. First Lunar Rover
Apollo 16	Apr 16 to 27, 1972	265 hrs 51 min 5 sec	1.5	64	Young Mattingly Duke	CM 113 *Casper* LM 11 *Orion*	Saturn 511. Explore Descartes region
Apollo 17	Dec. 7 to 17, 1972	301 hrs 51 min 59 sec	2	75	Cernan Evans Schmitt	CM 114 *America* LM 12 *Challenger*	Saturn 512. Explore Taurus-Littow area. Only Saturn V night launch
Skylab 1	May 14, 1973, to July 11 1979	?	—	—	unmanned	Saturn S-IVB shell, converted to manned workshop	Saturn 513, 2-stage configuration. Orbiting space workshop for scientific experimentation. Orbital altitude 271 nautical miles
Skylab 2	May 25 to Jun 22, 1973 (28 days)	672 hrs 49 min 49 sec	404	—	Conrad Weitz Kerwin	CM 116	Saturn SA-206. First Skylab occupancy
Skylab 3	Jul 28 to Sep 25, 1973 (59.5 days)	1427 hrs 9 min 4 sec	858	—	Bean Lousma Garriott	CM 117	Saturn SA-207. Second Skylab crew
Skylab 4	Nov. 10, 1973 to Feb 8, 1974 (84 days)	2017 hrs 16 min 30 sec	1214	—	Carr Pogue Gibson	CM 118	Saturn SA-208. Third Skylab crew
Apollo-Soyuz Test Project	Jul 15 to 24, 1975	217 hrs 30 min	138	—	Stafford Brand Slayton Leonov	CM 111 DM 2 (Docking Module)	Saturn SA-210. International cooperation in space. Craft remained docked for two days

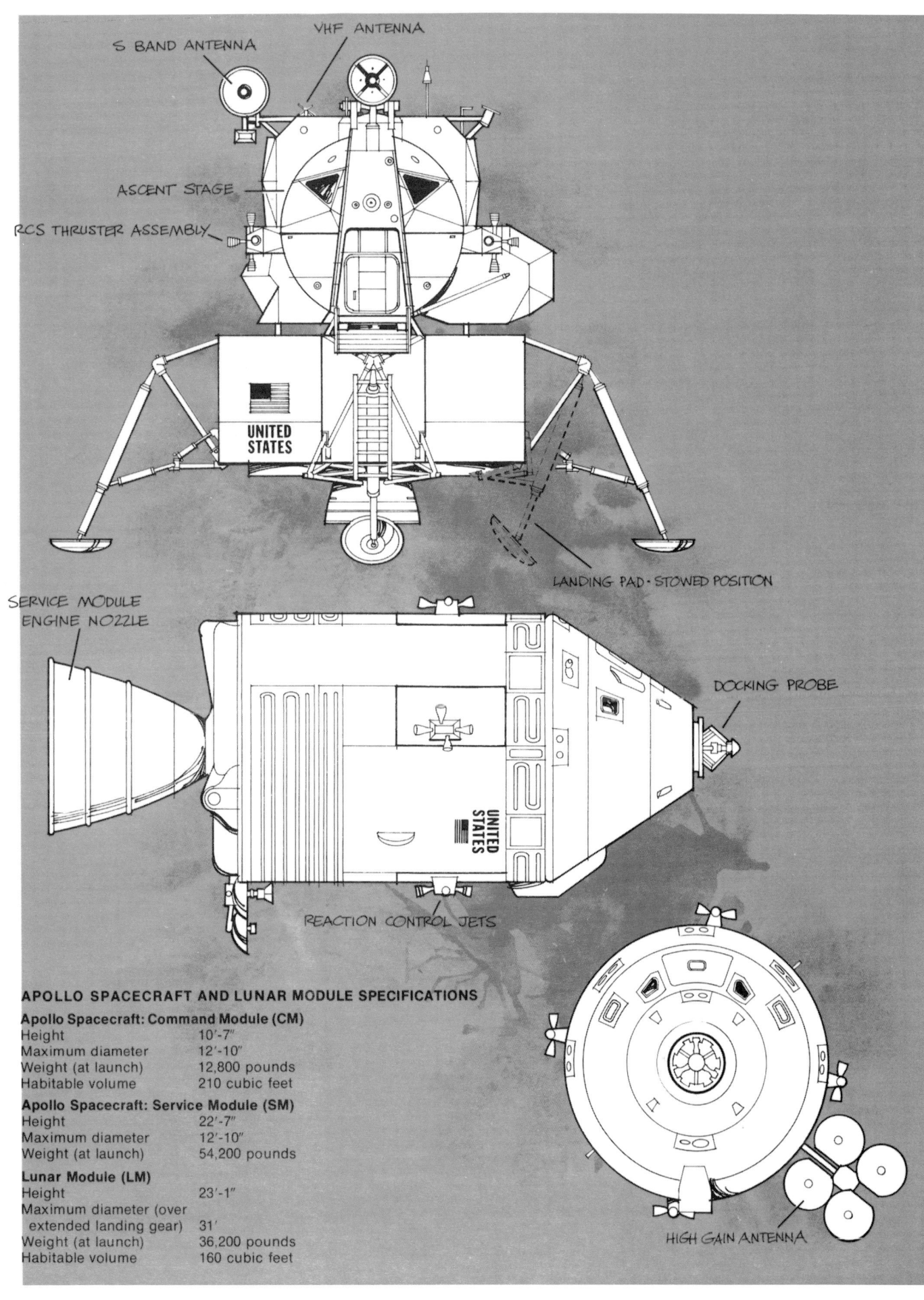

APOLLO SPACECRAFT AND LUNAR MODULE SPECIFICATIONS

Apollo Spacecraft: Command Module (CM)
Height	10'-7"
Maximum diameter	12'-10"
Weight (at launch)	12,800 pounds
Habitable volume	210 cubic feet

Apollo Spacecraft: Service Module (SM)
Height	22'-7"
Maximum diameter	12'-10"
Weight (at launch)	54,200 pounds

Lunar Module (LM)
Height	23'-1"
Maximum diameter (over extended landing gear)	31'
Weight (at launch)	36,200 pounds
Habitable volume	160 cubic feet

its engines to change course and return to earth.

If such an explosion had taken place during the Apollo 8 moon flight (when no Lunar Module was attached), or on any lunar landing mission during its return to earth from the moon (when the Lunar Module had already been jettisoned), the astronauts would have perished. The accident was traced to a test procedure at the launch site, which included a voltage application to heaters in the liquid oxygen tank. The power levels for the heaters had been decreased between Apollo 12 and Apollo 13, but one test group never got the word, and used the old current levels, overloading the heating element and burning off its insulation. In flight, the bare wire glowed white hot inside the tank, setting off an explosion as soon as it was turned on.

Safety considerations were high during Skylab, when the Apollo Command/Service modules had to remain docked to the space station for up to three months. A rescue Apollo was also available, to be flown by a two-man astronaut crew. Food lockers behind the crew couches were removed so that two additional seats could be installed, allowing a two-man Apollo team to launch, dock with Skylab, and pick up the three men in orbit. All five would return to earth inside the single Apollo. A similar design was possible during the Apollo-Soyuz linkup in the event that the two Russian cosmonauts had to return to earth along with the three Americans.

Once, the rescue ship was almost launched, when leaking fuel from the second Apollo visit to Skylab in August 1973 seemed to indicate that the spacecraft had become disabled. Ground controllers determined that it was still safe to fly. But a rescue ship would not have helped the third and last crew when, during the return to earth, the retro-rocket control system failed, stranding the astronauts in orbit (as in the prophetic movie *Marooned*). Fortunately, an alternate procedure on the next orbit allowed the men to come home safely, there never would have been time to launch the rescue ship before the marooned Skylab astronauts suffocated.

The various modules of the Apollo spacecraft were built and tested all over the United States. From Los Angeles came the Command and Service modules, built by North American Rockwell. From Long Island came the Lunar Module, built by Grumman Aerospace Corporation. The separate sections were transported by air, inside a special airplane with an enlarged cargo hold nicknamed the "Pregnant Guppy."

The design of a manned moonship had occupied the imaginations of space-minded engineers for decades before Apollo was born as the child of the Cold War competition between the U.S.A. and U.S.S.R. that fostered the so-called "space race." Years before President John F. Kennedy committed America to the moon landing "within this decade" (he made the speech in May 1961, a month after the Russians put the first man into space), space theorists in Russia and England had dreamed of a squat, unstreamlined "moon bug" that would land the first men on the moon. When the Apollo 11 crew landed on July 20, 1969, it was in a spaceship which bore an uncanny resemblance to designs published by the British Interplanetary Society—in 1939! And the technique of sending a scout ship down to the moon from an orbiting mother ship had been proposed a generation before by a Russian mathematician, but the idea was forgotten in his own country and had to be reinvented in the United States.

Most science fiction scenarios of moon flights had foreseen a winged spaceship blasting off from earth, touching down on the moon, and then returning directly to earth. When space engineers, responding to President Kennedy's rallying cry, actually plotted the fuel requirements and rocket power needed, they ran into major problems.

The moonbound rocket was much too heavy. Either it would have to be assembled from hardware launched separately into earth orbit, or a giant rocket twice the size of the still-imaginary Saturn V would be needed. The fuel cost of taking all of the main equipment and earth-landing hardware down to the moon's surface and back into space was simply prohibitive.

A young NASA engineer, John Houbolt, found the shortcut, unknowingly duplicating the logic of the forgotten Russian, Yuri Kondryatuk. The main spaceship could circle the moon at a safe altitude while a scout ship made the actual landing. The total weight of such a spaceship combination would only be half as much as the direct-landing-and-return system.

Houbolt's plan met strong resistance, since it required the development of an entirely new space vehicle, and it required the linkup of the moon landing vehicle with the mother ship on the way back to earth. These feats were untried, and might prove impossible. But when former critic Dr. Wernher von Braun studied Houbolt's idea and decided he had been wrong in opposing it, the American space team adopted it enthusiastically.

The author's lunar module looks great on an ersatz moon surface. David highly modified the Monogram lunar lander to depict *Eagle*.

"THE EAGLE HAS LANDED!"

A lunar module made from the 1/48 scale Monogram kit

DAVID SENECHAL

For those old enough to remember (I was 12), July 20, 1969 is a never-to-be-forgotten date. That's the day Neil Armstrong and Buzz Aldrin flew their lunar module to the first manned lunar landing. Twelve lunar modules (LMs) were virtually hand built at the Grumman Corporation in Bethpage, New York. Although similar, each was unique—improvements and increasing mission duration and complexity produced detail changes.

Because of its historical significance, I chose to model LM-5, *Eagle*, which was used on the Apollo 11 mission. Many kits released over the years have purportedly been the Apollo 11 LM, but they've all fallen short on accuracy. The closest was the 1/48 scale Monogram kit, which I used as the starting point for my replica of the LM-5.

My research for this project took place over three years. It involved careful study of hundreds of photographs—virtually all of the photos of LM-5 taken on the moon, as well as many of those taken by Grumman during its construction. It also included examination of two complete LMs still in existence.

LM-9, a "flight-ready" vehicle, is located at the Kennedy Space Center at Cape Canaveral, Florida. It is a later version, and wasn't much help in modeling LM-5. LM-2 is at the Smithsonian's Air & Space Museum in Washington, D.C., refurbished to be a duplicate of LM-5, and at first glance, looks correct. Close inspection, however, revealed many inaccuracies that made this LM too not much help. Both, though, provided a feel for what a lunar module is like, and supplied useful details of the landing gear, antennas, and the materials used in their construction.

Materials and color notes. The overviews give you a look at the completed LM-5 and some of its components. Most of the ascent (upper) stage of the real lunar module was covered with sheet aluminum, which acted as a micrometeoroid shield to protect the crew. I represented this with Testor's gloss silver paint. Areas subject to high temperatures were covered with a black material consisting of thin nickel foil and Inconel. I duplicated this surface by using either Floquil or Testor's flat black. The paint was applied either directly to the plastic or to household aluminum foil which was then attached to the plastic. Several areas were covered with a medium gold aluminized Mylar; for these, I used the wrinkled gold foil wrappers from Fanny Farmer chocolates.

One area required the use of actual aluminized Mylar. Several antenna supports and parts of the landing gear had a gold polymide tape holding thermal insulation in place. I cut ¼"-wide strips from a Mylar "thermal" emergency blanket to produce this effect. It was the right color and had the strength to be wrapped and pulled tight. All foil and Mylar on the model was attached using Microscale Metal Foil Adhesive.

Most of the descent, or lower, stage was covered with aluminized Mylar or the nickel/Inconel material. Additional areas were covered by a dark gold substance called H-film. H-film is similar to Mylar, but made to withstand higher temperatures. I used a darker gold candy wrapper foil here. I think I gained ten pounds while constructing this model!

Ascent stage. Monogram's ascent stage is divided into three sections: front, middle, and rear. I built them separately, then joined them after all three sections were complete. Many of the scratchbuilt parts were so fragile that they could

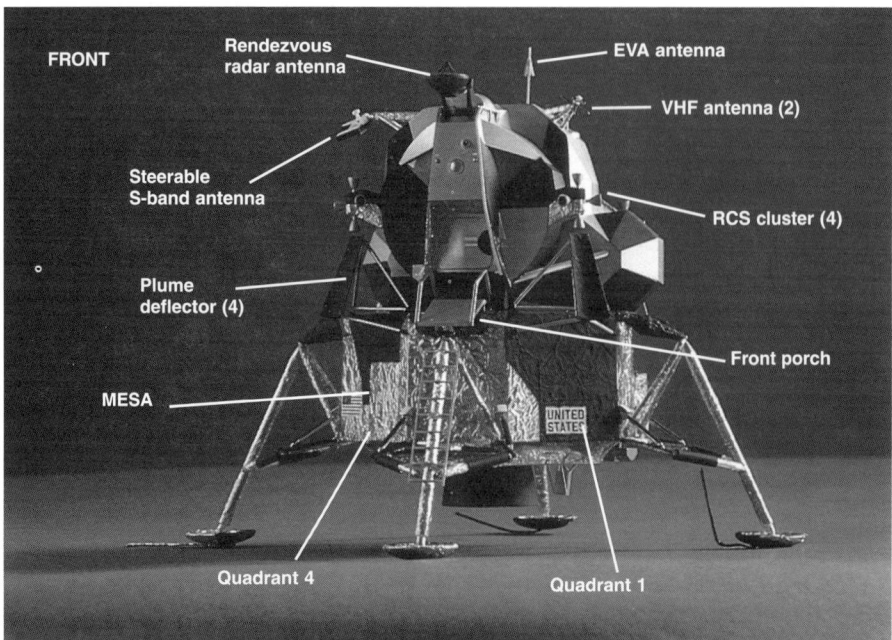

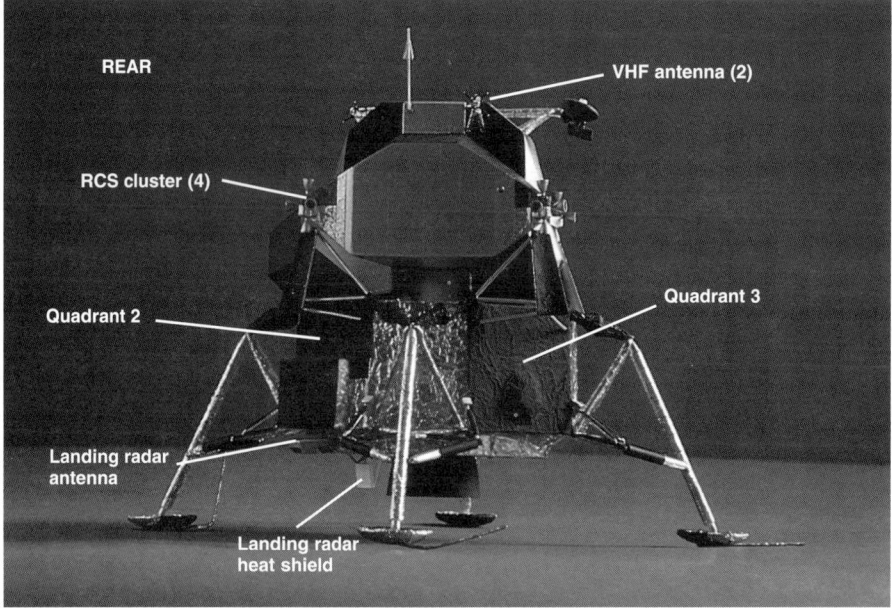

tolerate only a minimum of handling.

I started with the middle section, filling and sanding so the parts fit properly. I scribed panel lines on each side, then sanded away the raised panel sections on the upper back. The antenna placement holes on the top were filled with short pieces of sprue and putty, then new holes were drilled.

The antennas supplied with the kits were incomplete and out of scale, so I made new ones. I made new supports for the VHF antennas from .047" plastic rod. After assembly, I covered them with thin silver foil, then wrapped them with Mylar strips. The antennas themselves were made from three diameters of plastic rod, .020" brass rod, and thin wire taken from the armature of a 79-cent Radio Shack electric motor. I mounted the antennas on their supports using Duro Quick Gel Super Glue.

The kit's supports for the steerable S-band antenna were usable after minor alterations to make them fit their new holes, but the dish and associated hardware required additional scratchbuilt details (Fig. 1). After finishing, I put the antennas aside until the rest of the middle section was finished.

Photographed by Apollo 11 commander Neil Armstrong, lunar module pilot Edwin "Buzz" Aldrin Jr. deploys the Passive Seismic Experiments Packet (PSEP) on the lunar surface. NASA Houston No. AS11-40-5949.

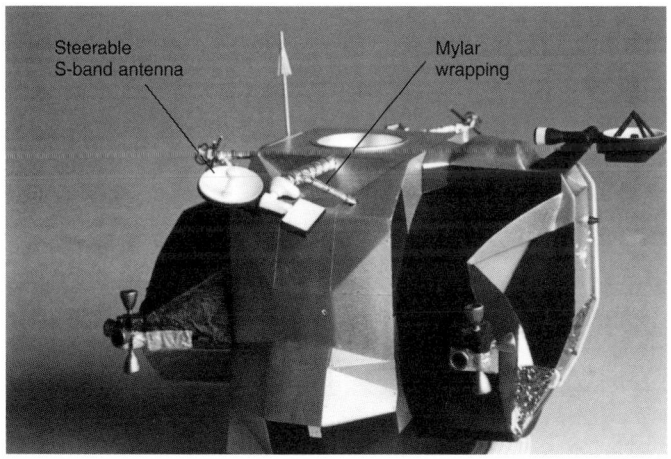

Fig. 1

The ascent stage displays an intricate pattern of black and silver finishes. I started with black since silver paint covers well but is difficult to mask over. Even though most of the surface is silver, I painted the entire assembly black to avoid any surface irregularity when the silver was applied.

The middle section was completed by painting the ascent engine skirt flat white, installing the various antennas, and applying dark gold foil to the bottom (Fig. 2).

The rear section involved the first large-scale use of J foil. First, I cemented the parts together (excluding the RCS clusters) and filled the seams. I sanded off the raised panel lines on the back and scribed new ones in the proper locations. The left and right ends were then covered with thin silver foil. Next, I constructed the EVA

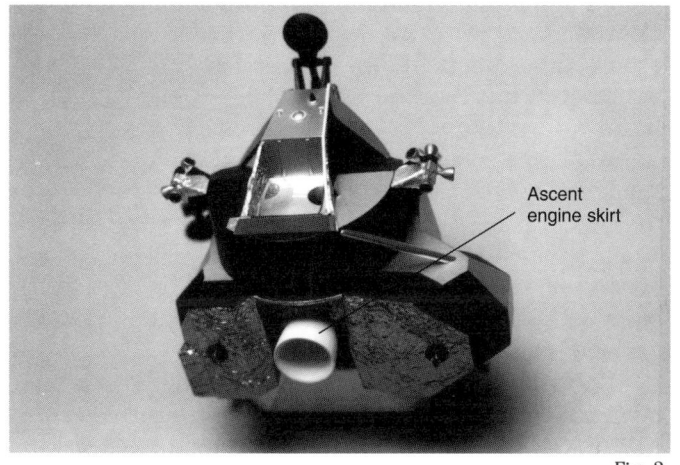

Fig. 2

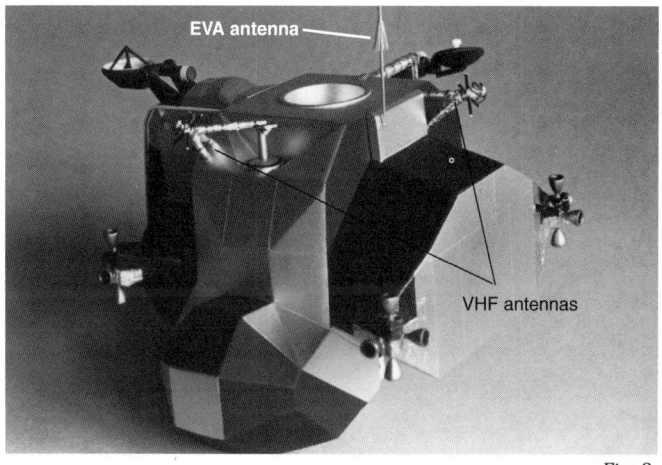

Fig. 3

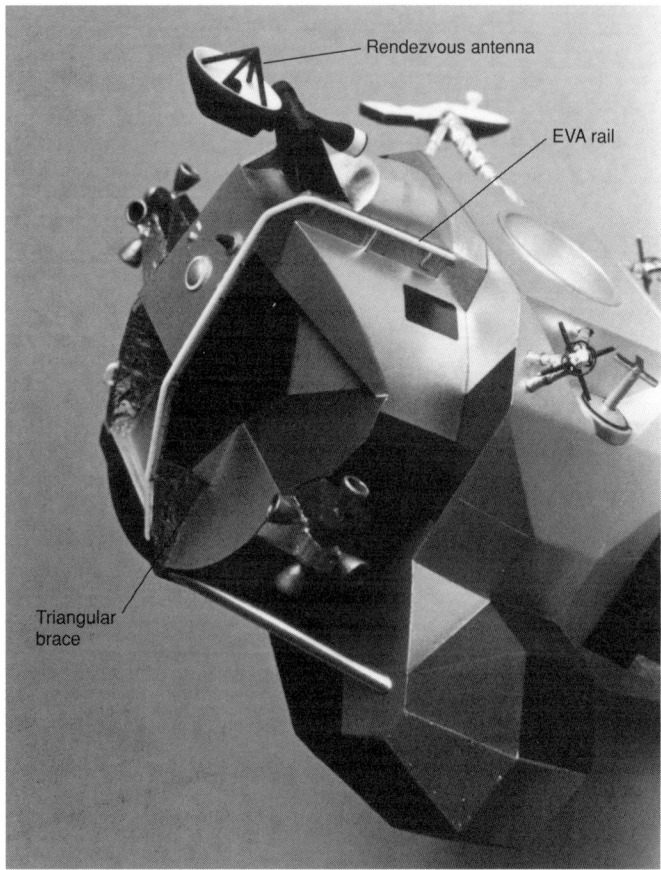

Fig. 4

Fig. 5

antenna using .025" plastic rod as a mast, and more thin Radio Shack wire for the nine "radials" (Fig. 3). This antenna was attached in the unstowed (vertical) position, and the rear section painted. I left the foil unpainted until after I added the RCS clusters.

I used four progressively larger drill bits to hollow out the RCS thruster skirts, then covered the remaining areas of these pieces with foil. Irregularly shaped foil "flaps" helped blend the RCS clusters in with the rest of the rear section once they were installed. I painted the foil with silver and black, and the RCS engine skirts with a mixture of gold and silver.

The front section of the ascent stage was the most difficult to build (Fig. 4). First, I cut the rendezvous antenna from its mount and completely rebuilt it. Next, I trimmed away the subreflector assembly inside the antenna dish and shortened both the shaft and counterweight by 2 mm. The angular protrusion on the rear of the dish was filled using small pieces of plastic and putty to achieve a smooth shape. I painted the inside of the dish flat white, and the rest of the antenna assembly flat black. I built a new subreflector assembly and attached it with white glue after the front section was painted.

I fabricated the two EVA rails on either side of the ascent stage's "face" from .025" plastic rod and .020" x .030" plastic strip. The triangular braces beneath the windows were made from .010" sheet styrene and attached. Now I could paint the front section; again, black first, then silver. The black areas on either side of the "face" were made from precut and painted aluminum foil (Fig. 5). I applied small sections of gold foil around the hatch. The seams were camouflaged by using an irregular edge to blend with the wrinkles of gold foil.

To make the windows, I used clear sheet plastic with the inside surfaces painted light silver-blue (actual LM windows had a blue thermal-reflective coating). Attaching and finishing the RCS clusters completed the front section.

I carefully joined the three completed sections and added a tubular brace between the front and middle sections. The kit's brace was not to scale and had an incorrect taper, so I made a replacement from plastic rod from my spare parts box.

Descent stage. The model's descent stage has virtually no exposed surfaces; it's covered with aluminum foil. Since this was a new technique for me, I used an old Monogram LM for practice—it serves as a "mannequin" for cutting and fitting the individual pieces of foil before applying them to the good model. The main octagonal piece required modification to match the actual LM. I cut a rectangular section from quadrant 4 (see the overviews for position of the quadrants); this is the location of the Modular Equipment Storage Area (MESA). The cylindrical bulge on quadrant 1 was also removed and the resulting gap filled with sheet styrene. I cemented the top and bottom pieces into place, but both pieces required trimming to achieve a good fit.

Work on the landing gear came next (Fig. 6). First, I sanded the secondary struts to provide a good fit for the sections of ⅛" aluminum tubing I placed over them. I painted these assemblies flat black with silver tips and light gray where the lower gear assembly meets the main body. This virtually completed the painting of the descent stage.

Covering the partially complete landing gear came next. I filled the tips of the upper outrigger assembly with putty, leaving a rough, uneven texture, then covered them with unpainted aluminum foil. Finishing the

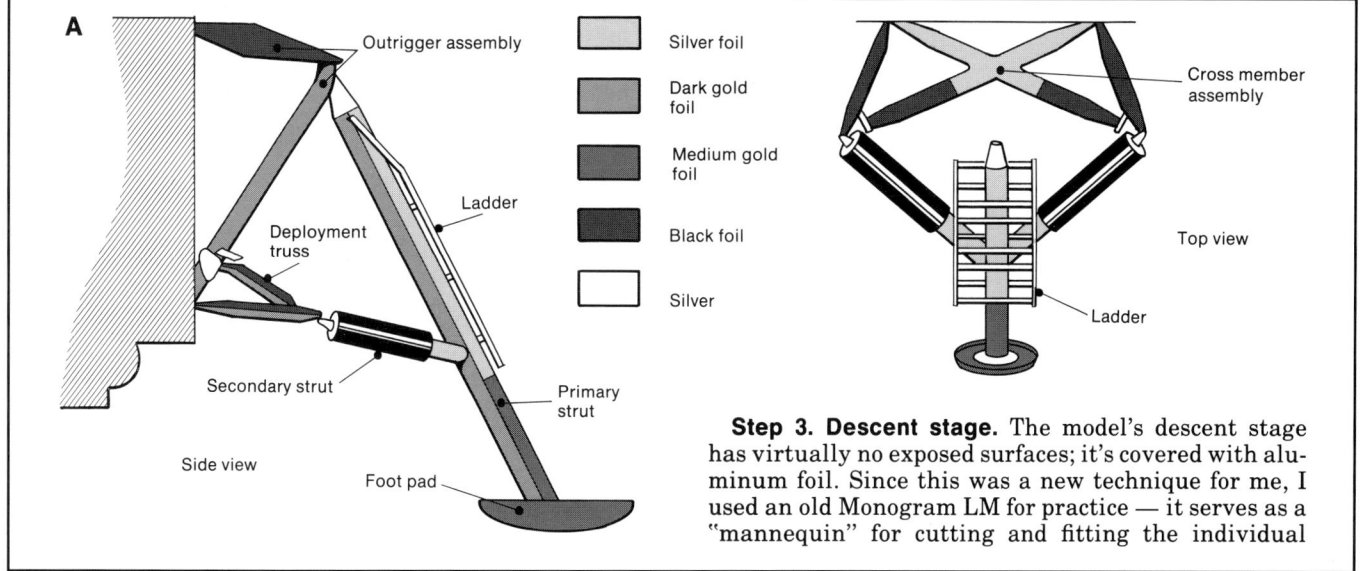

Step 3. Descent stage. The model's descent stage has virtually no exposed surfaces; it's covered with aluminum foil. Since this was a new technique for me, I used an old Monogram LM for practice — it serves as a "mannequin" for cutting and fitting the individual

Fig. 6

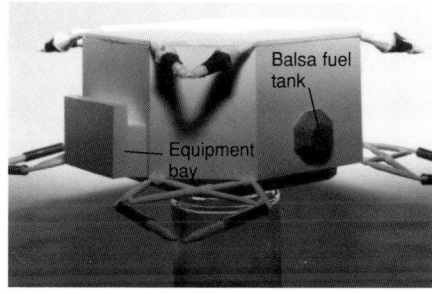

Fig. 7

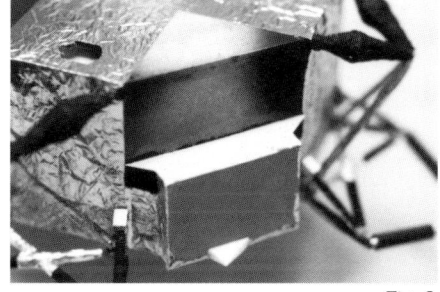

Fig. 8

Fig. 9

secondary strut assembly required quite a bit of time since the upper and lower halves of each brace and strut were covered with three different colors of foil. Since the struts were tapered at either end, the foil had to be cut in a "cigar" shape to fit properly.

Next, I scratchbuilt the rear quadrants. I added a balsa octagonal fuel tank to quadrant 3 (Fig. 7), then an equipment bay and landing radar antenna shield 4 made of balsa, .010" sheet styrene, and putty to quadrant 2.

I painted the descent stage with medium gold followed by dark gold, (Fig. 8). Next, I painted the lower outrigger struts silver, covered them with dark gold foil, and installed them. Then I added two scratchbuilt ascent stage support struts and covered them with foil.

I covered the black areas of the descent stage with aluminum foil, cut and fitted prior to painting, then cemented into place. I covered the tips of the upper outrigger assemblies with foil and painted them to match the adjacent black areas. The MESA was constructed of sheet styrene, covered with medium gold foil and inserted into the already-cut opening (Fig. 9).

The home stretch. Just a few details remain. First, I sanded off the ring-like protrusions on the kit's primary landing gear struts. I painted the struts silver, then covered them with silver and medium gold foil on the outboard edges, and dark gold foil on the inboard edges. Wrapping the struts with bands of Mylar finishes the struts.

The ladder supplied in the kit was too bulky and had too many rungs. I 4 made a new one with .025" rod and .020" x .030" styrene strips (Fig. 10), painted it silver, then mounted it on the forward strut with gel-type super glue. The attachment points were covered with foil, then wrapped with Mylar strips. The foot pads required trimming

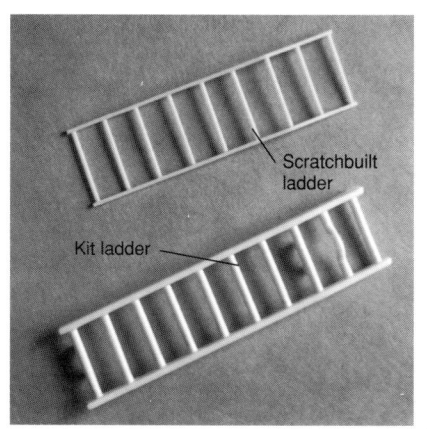

Fig. 10

Fig. 11

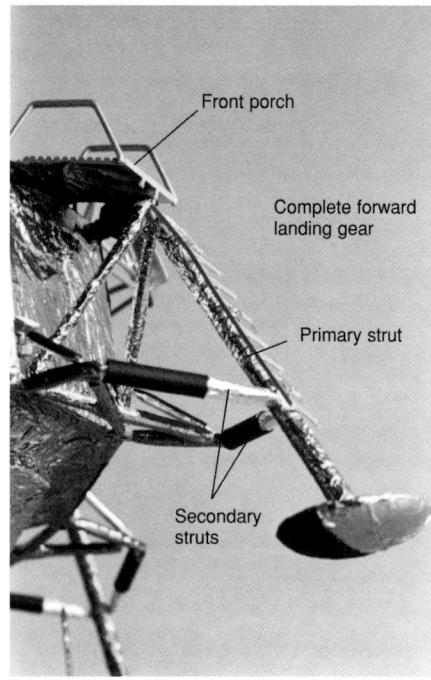

Fig. 12

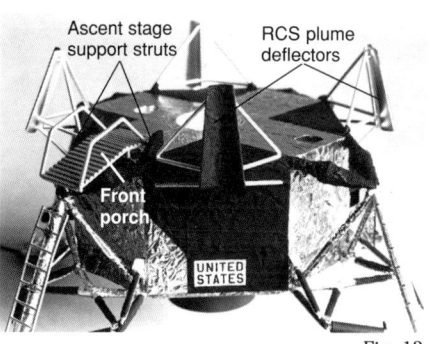

Fig. 13

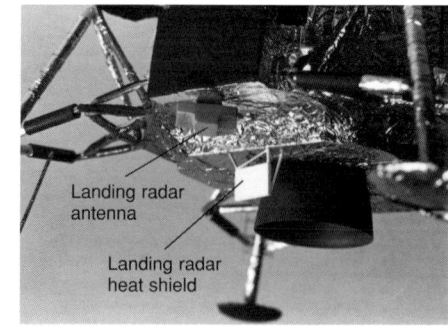

Fig. 14

and drilling before gold and black foil was applied (Figs. 11 and 12). Three of the four pads had landing probes, which I made from .025" rod, covered with two-tone foil, and super glued into the pre-drilled holes on the pads. On the actual LM, these were designed to crush or bend on impact with the lunar surface (LM-5's probes splayed to the right at various angles).

Although the Monogram front porch was designed to be installed backwards, it was usable after alterations. I removed the circular protrusions and sanded the porch to half its original thickness.

I attached the rails directly to the sides, then built and installed the mounting brackets (Fig. 13). The distinctive color of the porch was duplicated by using equal amounts of flat brown, flat tan, and gold paints.

The RCS plume deflector panels came next. The kit's panels were sanded to half their original thickness, joined with their respective support rods, painted silver, and the fronts covered with black foil. I attached the four completed deflectors to the descent stage with super glue and blended the joints with black foil.

Final assembly involved scratchbuilding and attaching the landing radar antenna and landing radar heat shield (Fig. 14). I painted the descent engine skirt with an equal mix of flat gray and flat black paint then cemented it into place. I applied the kit decals, and finally super glued the ascent and decent stages together.

Constructing my lunar module took nearly 200 hours. Although scratchbuilding and modifying took a great deal of time, the majority was spent on the tedious application of the fragile foil. Handling the foil was tricky, too—once attached, further touching would burnish the foil to the plastic, destroying the wrinkle effect. It's as fragile as the real thing! I recommend mounting the model to a base so you won't have to touch it.

My model is now safely under glass, and on permanent display in the headquarters building of the Grumman Corporation in Bethpage, New York. Many thanks to Lois Lovisolo, Grumman Corporation historian, and Josh Stoff, space curator, Cradle of Aviation Museum, Garden City, New York. Without their help, research for this project would have been nearly impossible.

SOURCES

• Styrene sheet and rod: Evergreen Scale Models, 12808 N.E. 125th Way, Kirkland, WA 98034
• Mylar emergency thermal blanket: Metalized Products, Inc., 37 East Street, Winchester, MA 01890
• Electric motor (for wire): Radio Shack stock No. 273-223
• Stucco filler putty: Verlinden, Letterman & Stok, West Port Industrial Park, 804 Fee Fee Road, Maryland Heights, MO 63043

REFERENCES

• *Life,* "To the Moon and Back," 1969 (special edition)
• *Life in Space,* Time-Life Books, Chicago, 1983
• *National Geographic,* December 1969
• Smithsonian National Air and Space Museum Archival Videodisc V (the NASA collection), Washington, D.C., 1987

Plastic shrouds protected Apollo 11 thrusters from salty, humid Florida atmosphere.

SATURN V

America's unparalleled moon rocket

JAMES E. OBERG

Superlatives fail when it comes to describing the Saturn V moon rocket. It was the biggest space booster in the world, and bigger ones may never be built. Combined with its awesome power was delicate precision and solid reliability—all attributes necessary for the rocket which was to send men to the moon.

Space flight offers many contrasts: Months before launch date, the Saturn V for Apollo 10 that carried astronauts Stafford, Young, and Cernan at speeds exceeding 25,000 mph was carried to the launch complex at speeds no faster than a slow walk. Roadbed is 7 feet thick.

Five engines that produced a total maximum thrust of 7,500,000 pounds lifted Apollo 15 space vehicle from Pad A, Launch Complex 39, Kennedy Space Center, Florida, at 9:34:00. 79 a.m. (EDT), July 26, 1971.

Imagine a machine with 40 million horsepower. Imagine a machine that guzzled fuel faster than all operating American automobiles combined. Imagine a machine the very sound of which could smash unprotected human bodies for hundreds of yards around. Imagine a machine that took 10,000 man-years to build but was used for only 10 minutes. That's Saturn V.

Fifteen Saturn V rockets were built during the 1960s for the Apollo moon program. After two unmanned flights in 1967 and 1968, Saturn Vs carried 10 manned Apollo expeditions in the years 1968–1972. One Saturn V placed the giant Skylab space station in orbit in 1973, and the remaining two served as backup vehicles and then were placed on exhibit in museums (actually, of course, outside the museums!).

The Saturn V was the largest member of the Saturn series of rockets, which were conceived by the rocket team of Wernher von Braun at the Huntsville, Alabama, Redstone Arsenal (later, under the National Aeronautics and Space Administration, it became the George C. Marshall Space Flight Center). The smaller Saturn I began testing in 1961 and launched test Apollos into earth orbit. The manned Apollo 7, the three Skylab visits, and the Apollo-Soyuz linkup were launched atop the Saturn I, which had about one-fifth the power of the Saturn V

A Saturn V rocket stood 363 feet high. It was 33 feet wide at its base, excluding fins. It had three stages, of which only the third stage went into orbit with the Apollo spaceship. After a few hours of navigational checks, the third stage was fired once again to

(Above) Giant Crawler/Transporter carried Saturn V for Apollo 15 and the Launcher Umbilical Tower from Vehicle Assembly Building to the Launch Complex. An ambitious modeler could build these structures from styrene. (Right) Fuel was delivered through piping from the tower.

propel the Apollo outward and upward towards the moon.

STAGE 1 (S-I C): For the first two minutes of flight, the Saturn was lifted by five giant F-1 rocket engines, the most powerful ever built anywhere in the world. Kerosene and liquid oxygen were burned at a rate of 10,000 pounds per second. Its job done, the first stage fell back to earth and tore apart in the atmosphere, with pieces splashing into the Atlantic 300 miles from Cape Canaveral, Florida.

STAGE 2 (S-II): Five J-2 engines pushed the remaining vehicle higher into space, using a high-efficiency propellant supply based on liquid oxygen and liquid hydrogen. This stage burned for three minutes before falling back to earth and disintegrating in the upper atmosphere 5,000 miles from the launch site.

STAGE 3 (S-IVB): A single J-2 engine powered the third and final rocket stage. Atop this stage was the computer unit which provided guidance to the entire three-stage rocket. This stage entered earth orbit, after which the rocket was shut down, usually with about half its fuel remaining. The third stage stayed in orbit several hours while careful navigational observations were made by the Apollo crew, the third-stage computer, and Mission Control at the Manned Spacecraft Center (now the Johnson Space Center) at Houston, Texas. With updated targeting information loaded into the computer, the third stage was fired again for the final lunge into deep space. The separation of the Apollo spacecraft was next in the moon-flight sequence. No watery grave awaited the spent third stage; space controllers aimed it smack into the moon to create artificial "moonquakes" for scientific instruments to monitor (on the test missions, however, the third stage was simply sent into orbit around the sun).

The flight profile of the Saturn rocket was, in theory, a simple one. But since nothing assembled by human minds and hands ever seems to work perfectly, numerous

Nine ramps gave access to each level of Apollo 15 before launch.

problems did arise. Every single Saturn rocket was successful in its mission, but often it had to rely on redundancy and the "safety margin" that designers had built in.

The first test flight of the Saturn V from John F. Kennedy Space Center, Florida, on November 9, 1967, went almost flawlessly. The only minor casualty was TV newsman Walter Cronkite, whose television studio—hastily constructed at the press site 3 miles from the launch pad—fell over on him during the powerful blastoff. The rocket itself, in a bold "all stages live" gamble, worked just about as planned.

Not so the second unmanned test the following April. During second-stage flight, severe "pogo oscillations" tore pieces of the rocket loose. Two of the five second-stage engines failed, yet the flexible control computer tilted the remaining

Saturn 505 carried Command Module *Charlie Brown* and Lunar Module *Snoopy* into lunar orbit during Apollo 10.

three engines slightly and burned them for a minute longer, compensating for most of the loss. And although the third-stage rocket burned during the ascent, it failed to burn the second time after coasting through space for several hours.

Rocket engineers worked furiously to eliminate the problems. "Pogo," a fierce up-down oscillation, was caused by the surge of fuel in the feed lines—surging which initially was set off by rocket vibration but which snowballed when each surge fed extra fuel into a succeeding surge a second or two later.

Uncontrolled, Pogo could (and did) tear rockets apart. Special flow controllers and reservoirs were designed and installed in the next Saturn V. The J-2 failure was traced to vibration of fuel lines, in which metal fatigue caused small pipes to break off. The pipes were made more flexible.

The fixes apparently worked, because the Saturn Vs on the next five missions performed perfectly. Even when the rocket was hit by lightning a few minutes after launch during Apollo 12, the controller computers continued to guide the spaceship on its course.

At 13:13 Houston time, April 11, 1970, Apollo 13 blasted off. It was to take its place in the history books as the only moon flight that failed. But many hours before the explosion (see page 21) that nearly doomed the astronauts to a lingering space death, the Apollo 13 Saturn V suffered the most serious rocket failure of any American manned space shot. One of the five engines of the second stage failed, and the remaining engines in the second and third stages had to burn longer to compensate. Luckily, a fuel reserve had been included, and the launch toward the moon was successful. Fate, or just plain bad luck, clearly was out to get Apollo 13.

Five more Saturn Vs were subsequently launched, each carrying more and more weight as design improvements were added and safety margins were carefully reduced when they proved to be too generous. Lunar payloads increased from 87,382 pounds on Apollo 8 to 107,221 pounds on Apollo 17. The final Saturn V flight used only two stages to put Skylab into an earth orbit. The burned-out second stage remained in orbit for a year, trailing behind the space station but gradually slipping back to earth under the weak air drag 200 miles high. For a few days, observers on earth feared it might hit the ground intact and cause property damage or human casualties. But eventually it plunged harmlessly into the North Atlantic, out of sight.

Stages, souvenirs, and the Soviets.
The three stages of the Saturn plus the rocket engines and other components were built across the United States and shipped by barge, airplane, and truck to Kennedy

Space Center. Stage one was built by Boeing at Huntsville and Michoud, Louisiana. Stage two was built by North American at Seal Beach, California. Stage three was built by Douglas at Huntington Beach, California. Rocketdyne built all the engines at Canoga Park, in the same state.

Once the stages arrived in Florida, they were rolled on flatbed trailers into the giant Vehicle Assembly Building (the VAB). Standing 550 feet high, the VAB was the largest building in the world when it was constructed. Inside, four separate Saturn rockets could be assembled, each on its own movable pedestal in its own assembly bay. (In practice, only two of the bays were actually used, with the others in reserve in case of accident or additional launch requirements.) Once assembled, the three-stage Saturn V and its Apollo spaceship were transported vertically on the pedestal to the launch site 3 miles away.

A giant crawler tractor moved the pedestal carrying the unfueled Saturn along a dual-lane gravel road 130 feet wide at a speed of 3 mph. Once the pedestal with the Saturn was placed on the launch pad, the tractor slipped out from underneath and crept away to a safe distance, always ready to be called back in case the Saturn had to be returned to the VAB for repair, replacement, or refuge from hurricanes.

Launchings were eyewitnessed by tens of thousands of people from all over the country and by millions on live TV. Half a million Americans gathered in Florida in July 1969 to see the sixth Saturn V carry Apollo 11 astronauts Neil A. Armstrong, Michael Collins, and Edwin E. Aldrin Jr. into space. Crowds dwindled on later "routine" missions, with an upsurge for the final and most spectacular moonshot, Apollo 17. This twelfth Saturn V booster was the only one to be launched at night, and it flared like a giant torch in the midnight skies over the East Coast.

Saturn V rockets left scattered souvenirs across the face of the earth and moon. A German freighter was showered with metal fragments from the Apollo 11 launch, and stories filtered out of the jungles of South America and Africa about aluminum sheets that fell from the sky and were worshipped as gods by the natives. Five new craters were dug on the moon by Saturn V third stages hitting the surface at 6,000 mph and splattering fragments across hundreds of square miles of "lurain" (lunar terrain).

The white paint with black patterns on the Saturn V booster was as much a matter of tradition as an engineering requirement. Fins were lettered so that tracking films could tell which way the booster twisted, while the general pattern was helpful in visual tracking. Mainly, however, the livery was a carryover from the V-2, Viking, Redstone, and Jupiter rocket days. The Saturn V was painted white and black because that was the way Wernher von Braun liked it (the paint layer weighed almost a ton and cost a few hundred pounds of payload capability, but the Saturn V could carry it without any problem).

Space operations in the Space Shuttle era do not require the weight-lifting power of the old Saturn Vs. Instead, space payloads can be built in smaller modules and brought into orbit separately, then linked together. The total payload

---

Apollo Booster Specifications: Saturn 1B

Height (with Apollo)	224'
Maximum diameter	21'7"
Weight (flight-ready)	1,300,000 pounds
Stages	2
Propellants	
1st stage	RP1 (kerosene), liquid oxygen
2nd stage	Liquid hydrogen, liquid oxygen
Maximum payload (low earth orbit)	40,000 pounds
Maximum 1st-stage thrust	1,600,000 pounds
Engines	
1st stage	8 H-1
2nd stage	1 J-2
Saturn 1B horsepower	24,000,000

Apollo Booster Specifications: Saturn V

Height (with Apollo)	363'
Maximum diameter	33'
Weight (flight-ready)	6,200,000 pounds
Stages	3
Propellants	
1st stage	RP1 (kerosene), liquid oxygen
2nd stage	1 J-2
3rd stage	liquid oxygen
Maximum payload (low earth orbit)	270,000 pounds
Maximum payload (trans-lunar injection)	100,000 pounds
Maximum 1st-stage thrust	7,500,000 pounds
Engines	
1st stage	5 F-1
2nd stage	5 J-2
3rd stage	1 J-2
Saturn 5 horsepower	160,000,000

Saturn V Construction Record

Total vehicles constructed	15
Unmanned flights	2
Manned Apollo missions	10
Skylab orbital booster	1
Backup units (later used as exhibits)	2

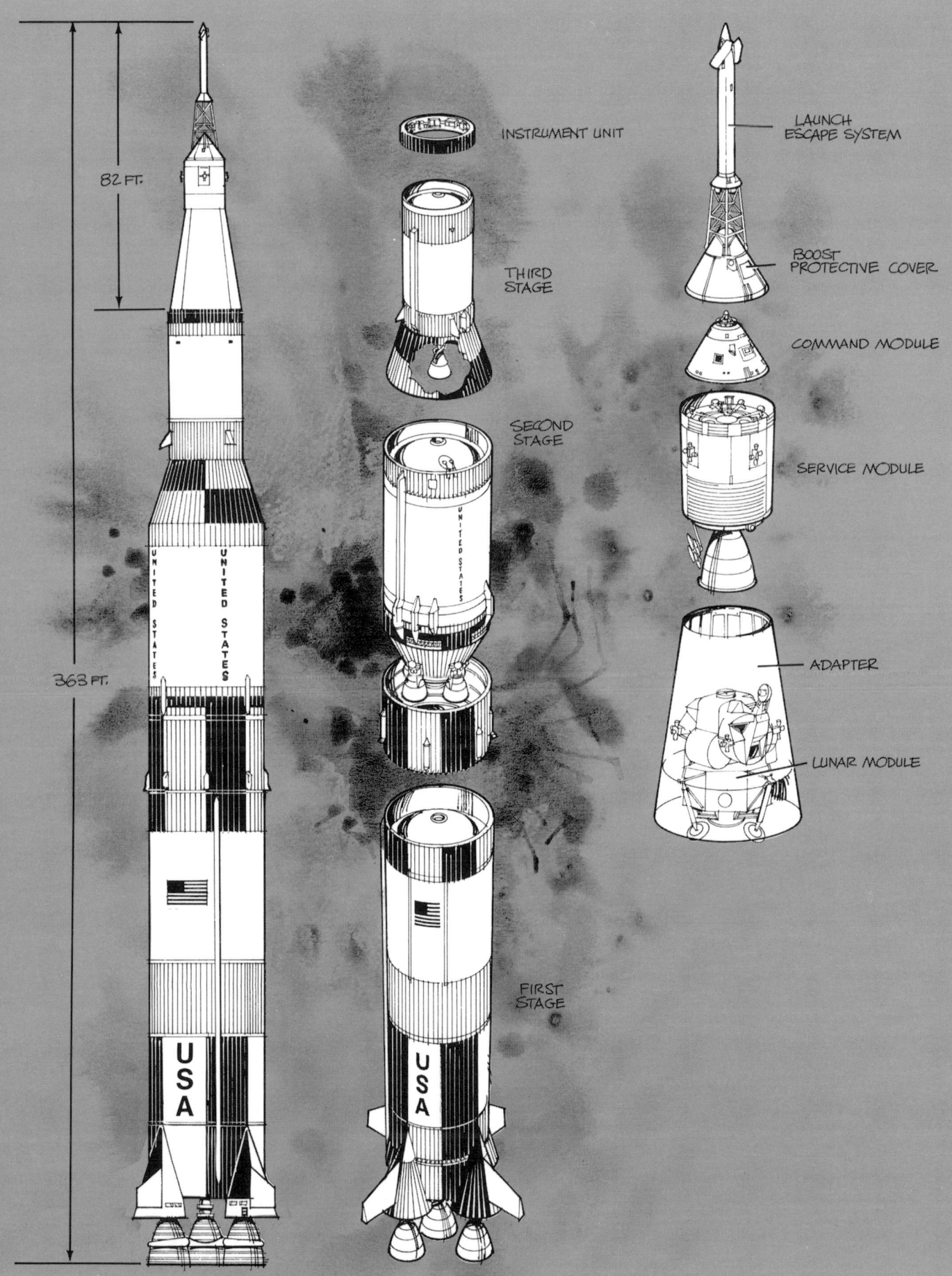

capacity of the Space Shuttle is about one-fourth that of a Saturn V, but at one-tenth the cost.

The seeming wastefulness of dropping multimilliondollar Saturn rockets into the oceans was a popular theme during the Apollo program, but critics did not realize what rocket designers had come to know: Throwing away the booster stages was actually the cheapest way to fly. Only major advances in space technology, spurred by the Apollo program, allowed space engineers to advance to the cheaper reusable Space Shuttle.

How can throwing away a booster be cheaper? The immense prices quoted for Saturn rockets (180 million dollars each) involved primarily the man-hour cost of checking out the rocket and preparing it for flight. The actual cost of the metal in the rocket was a small percentage of the final cost.

The engineers certainly did something right. They built an unprecedented space machine, using technologies never before tried (and many not even imagined when the project began), and they did it on schedule and under budget. The Saturn V never failed to deliver its payloads into space with the planned velocities and headings. And as their experience and confidence grew, these rocket builders added new and more powerful features to each subsequent Saturn V.

The American success must also be contrasted with the failure of the Soviet Union to build a similar rocket. Soviet space planners were equally committed to sending Russian cosmonauts to the moon, ahead of Apollo, but their technology base failed them and their program faltered. After the early Apollo successes, the Soviet man-to-the-moon program was canceled and covered up, and Moscow began a new propaganda tack of claiming that it had never really meant to send men to the moon, anyhow. This sour grapes whine was accepted in some Western circles because it coincided with old antispace prejudices, but the Soviets were lying.

In fact, a giant Soviet space booster was built with twice the liftoff thrust of the Saturn V but, because of inherent design deficiencies, with only about the same payload capacity. Despite reported launch attempts in the 1969–1972 period, the Soviet space rocket team was unable to carry off a successful flight. The secret rockets crashed in flames during the same years that Saturn Vs were blazing the American trails to the moon. It's a chapter of Soviet space history that Moscow never talks about. And it underscores the brilliant engineering work that made the Saturn V possible.

Saturn V's engines are silent now. Fragments carpet the Atlantic and the face of the moon. Surplus equipment lies on the ground at space museums at Johnson Space Center, Marshall Space Flight Center, and Kennedy Space Center. But even today, visitors to these rockets feel an incredible sense of awe at the size and power of the Saturn Vs, which in real life blazed paths across the skies and which burned their way into history books.

BEGINNER'S GUIDE TO PLASTIC KIT MODELING

HAROLD A. EDMONSON

Kit: Apollo-Saturn V
Manufacturer: Airfix, England, Kit No. 09170-5
Scale: 1/144
Dimensions: 30⅛" high (without display base); 5¼" diameter at fins; Display base—4¼" x 5⅞" x 1" high

The Airfix Saturn V is an ideal model for the beginner. The parts are large and easy to handle and assemble. The finer techniques of modeling—such as seamfilling and sanding—can be learned on the broad surfaces of the Saturn's stages. Spray-painting techniques can be developed using the Saturn. You might run into trouble with the masking and painting of ribbed surfaces, but it's just as well to acquire this modeling skill right from the beginning.

The tools I used for this project included a hobby knife (X-acto with No. 11 blade), a set of jeweler's files, wet-or-dry sandpaper (No. 400 and

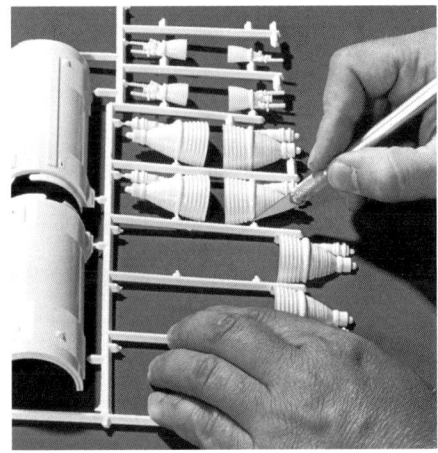
Fig. 1

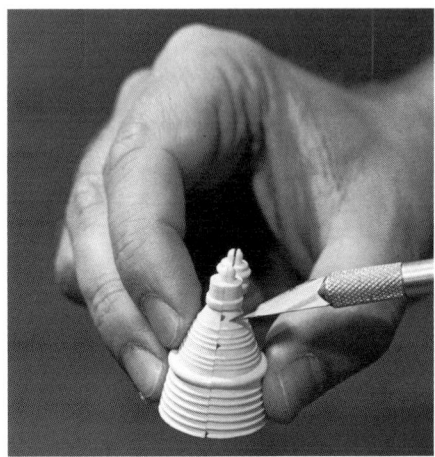
Fig. 2

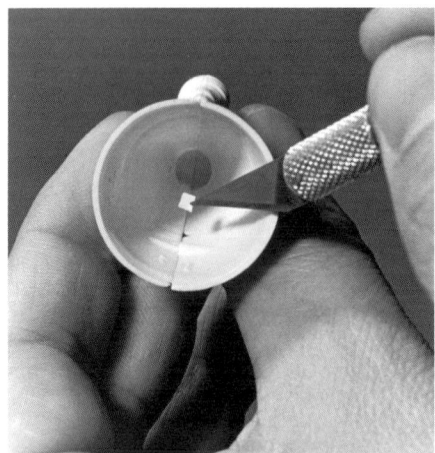
Fig. 3

No. 600), masking tape, Scotch Magic Transparent Tape, and liquid cement for plastics with a paint brush to use for application. Other supplies will be introduced during the construction process.

First, carefully read the instruction sheet. Look at the sprue tree and identify the parts; don't remove any parts until you understand where they will fit in the finished model. Remove parts only as they are needed for test-fitting, construction, or prior painting. Do not break the parts off the sprue tree with your fingers; you may lose a small chunk of the part on the sprue tree (Fig. 1). Cut the part away with your knife (or with a razor saw if the attaching plastic is very thick). You can always file or sand any portion of the sprue still attached to the part.

Test-fit all parts before doing any cementing. Make sure the pegs and tabs of one part fit the holes or slots of the next part. Use the hobby knife or files to trim and clean the parts until they fit snugly together.

Following the order of assembly shown in the instruction sheet, I test-fitted the lower part of the first stage, which included the stability fins and five F-1 rocket engines (Fig. 2). When I test-fitted the two rocket cone halves, I noted some small gaps or holes where the halves butted together—the plastic had not completely filled the mold when the part was made. This was one of the few faults I was able to find with the craftsmanship of the kit.

Using liquid cements, the first parts I cemented were the F-1 engines. After cementing all five, I noticed that the cones of several of the engines had at least one gap somewhere along the join line. The first step in correcting this problem was to fill in behind the gap with a small piece of plastic. Sheet plastic, available at most hobby shops, would be ideal for this purpose, as would flat pieces of scrap plastic.

I cut small squares of plastic from the styrene sheet and glued them inside the cone behind the gap (Fig. 3). The piece of plastic alone was enough to reasonably disguise the flaw, however, perfectionists would want to fill in the surface of the gap with a small dab of filler putty. Careful filing and sanding would then be required.

Although I tried to slightly flex the F-1 halves to align them perfectly, a ridge line nevertheless formed in certain places along the join line (Figs. 4 and 5). I used the tip of my flat file to remove some of this ridge and even used my round or rattail file to reach between the ribs of the cone.

The gaps in the cones of the F-1s led me to think about the second

Fig. 4

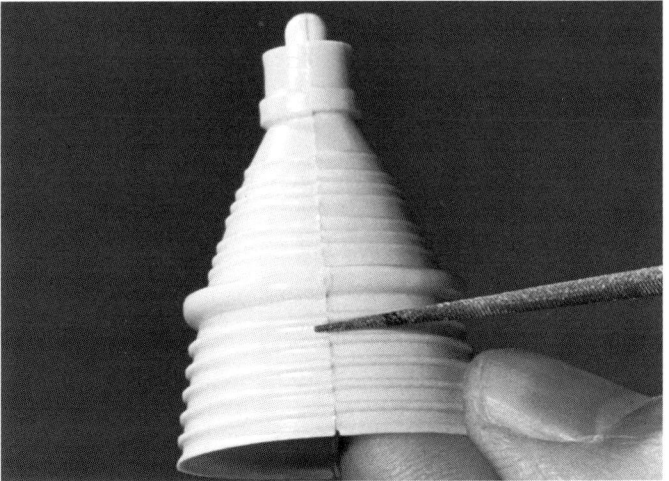
Fig. 5

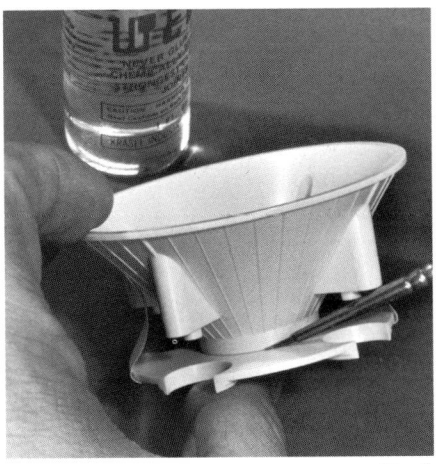
Fig. 6

Fig. 7

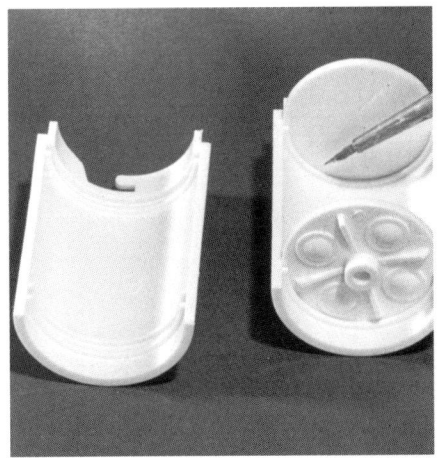
Fig. 8

stage J-2s. I jumped ahead of the construction sequence and test-fitted those halves. Happily, gaps did not appear. I went ahead and cemented them together because all the rocket engines would be painted silver and it would be easiest to paint them all at one time before assembly. Therefore, I completed assembly of all the rockets and attached them to a cardboard sheet to await painting.

I also placed the stability fins on the cardboard (later I'd use clothespins during painting). My test-fitting had indicated the fins would fit snugly without unsightly gaps requiring putty. Therefore, I could also paint them before assembly.

I resumed test-fitting by placing together the large cylindrical halves for stage one and stage two. Dome-like pieces represent the tops of fuel tanks and, in effect, form the tops of stage one and stage two (and also stage three). Because these pieces fit between two rims cast on the inside of the cylinders, they must be inserted before the halves can be cemented, and should be painted before any assembly is done.

Paint these domes yellow, which is the color of the actual cross-member assemblies and bracing visible at the top of stages one and two. An interesting project for the advanced modeler would be to scratchbuild the bracing that is actually found at the top of these stages, but for my purpose, I used the domes as intended in the kit.

On one piece of cardboard I placed the domes to be painted yellow. On another I began placing all intended for white paint. Before placing the domes, I filed and smoothed the circumference edge.

The rocket motors for stage one are installed in a flat engine-mounting plate that in turn is inserted as the bottom of stage one. Paint this plate white. Cone-shaped pieces are the engine mounting structures for stages two and three, and these fit inside rims just as the cylinder tops do. Therefore, the pieces must be installed at the time the stages are assembled and painted silver.

The second stage engine-mounting structure actually consists of a cone and a flat plate; I cemented them together at this time (Fig. 6). (This flat plate doesn't exist on the Saturn. A modeler who wants to omit it, however, will have to retain the center section as a support for the center rocket motor.)

Before I placed the mounting structures on my "silver" cardboard, I test-fit the J-2 rocket engines to them (Fig. 7). A molding seam showed on that surface area of the J-2 that would mate to the mounting structure, so I filed it flat.

By this time I was gathering a number of parts that would require silver paint, so I decided to see if I couldn't accomplish all the silver painting at one time. I studied the instruction sheet and gathered other silver parts. Some of the parts, such as the Service and Lunar modules, were in halves, so I assembled them at this time.

The cross members were inserted in one Service Module half and cemented in place (Fig. 8). When the glue was dry I glued the two halves together.

Next, I assembled the Lunar Module halves and trimmed and filed join lines where needed. The landing legs were cemented in place, and the LM joined the other parts on my "silver" cardboard.

Some of the remaining pieces included the Apollo capsule and escape rocket. The capsule would be painted silver, so after sanding some rough edges, I merely set it next to the other silver parts. The main heat shield would be painted a different color later. Since the capsule and heat shield ultimately would be cemented together, I saw no reason to use the three astronaut figures in my model.

Next I assembled the escape tower, including the cone that fits over the Apollo capsule, and placed it among the parts to be painted white.

Although I had three cardboard sheets holding the parts that would be painted silver, white, and yellow, I could have painted all the parts a flat white primer at this point to reveal imperfections in the parts. I didn't because I wanted to keep the painting steps as simple as possible. The yellow parts are so few and hidden as not to require special attention and I wanted to avoid paint buildup on the small and contoured silver parts. Later however when I painted the large stages I

29

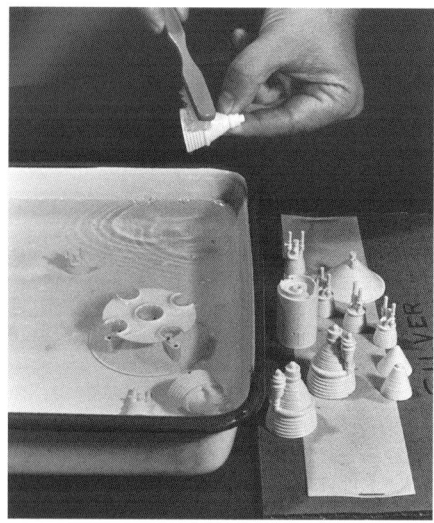

Fig. 9

Fig. 10

Fig. 11

did use a flat white as the first coat.

Before painting I washed the parts in dishsoapy water to remove all oils so the paint would adhere better (Fig. 9). Before painting, place the spray cans in a pan of warm water for several minutes to help the paint flow better.

I began spraying to the left of the objects and made a deliberate pass to the right, staying 9 to 12 inches away, and released the button only after my hand passed completely to the right of the objects (Fig. 10). Do not start spraying with the nozzle pointed at the object, and do not stop the motion or go too slowly while spraying: This will cause uneven coverage as well as pooling of the paint.

Because many of the silver parts were rounded, I painted them in several passes, rotating the cardboard between passes (Fig. 11). Because it's easy for too much paint to build up and begin running when you do this, I stopped completely at the halfway point and waited a day for the paint to dry before continuing. I was able, however, to paint both sides of the first-stage stability fins at one time by placing them in clothespin holders (Fig. 12).

Now, after painting, turn the aerosol cans upside down and press the nozzle until the paint clears and only propellant comes out. This prevents paint from clogging in the nozzle.

After the yellow and silver parts were dry, I was ready to assemble the stages and begin showing some progress! I cemented three cylinder domes and two rocket mounting structures in place in the halves of the major stages and then joined the other halves. Although I had prefit the halves as best I could, I still found it helpful and important to flex the halves into perfect alignment as the liquid cement was drying (Fig. 13).

Once the cement appeared to hold the halves in place as I wanted, I placed rubber bands tightly around the stages for overnight drying. In fact, I even placed some of the rubber bands in position at one end of the stage while I was still applying cement at the other end.

After allowing the stages to dry at least overnight, I used the edge of my knife blade to scrape along the join lines, taking care not to run my knife into any of the raised detail (Fig. 14). By and large, the seams looked smooth after adzing. But there were still a few gaps and uneven areas that could use putty.

I repeated the process I had used with the Service Module (Fig. 15): I applied thin coats of Green Putty cut with Testor's liquid cement (I used the Testor's bottle-cap brush as the applicator). I sanded with fine and then extra-fine Flex-I-Grit sanding film. After most

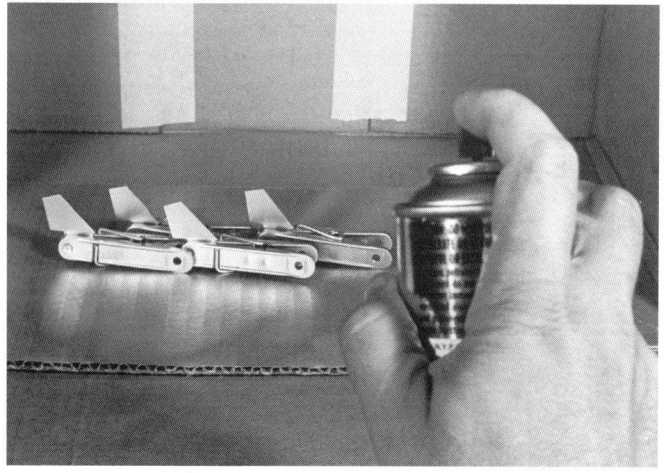

Fig. 12

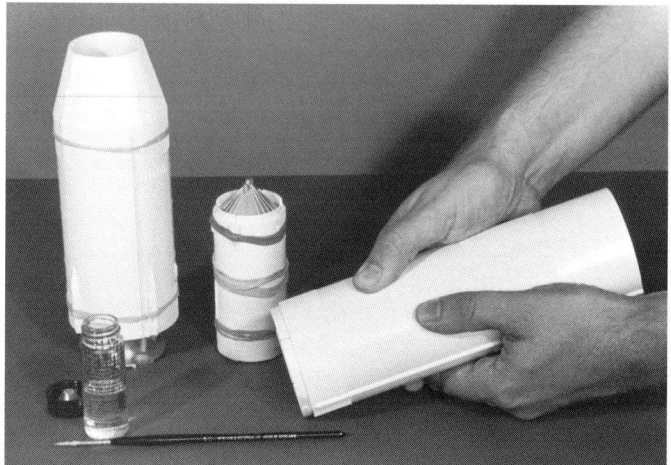

Fig. 13

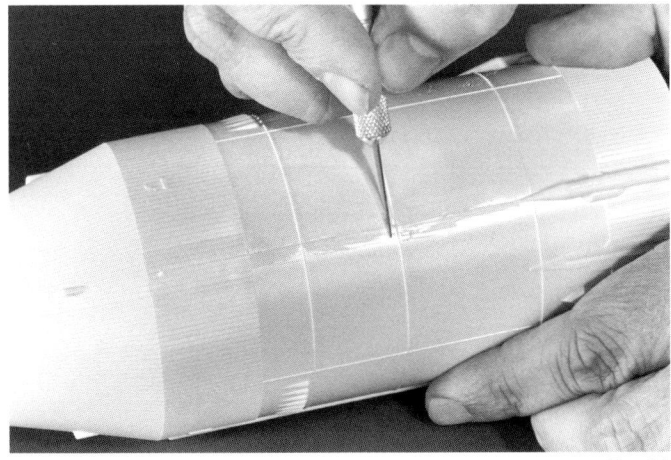
Fig. 14

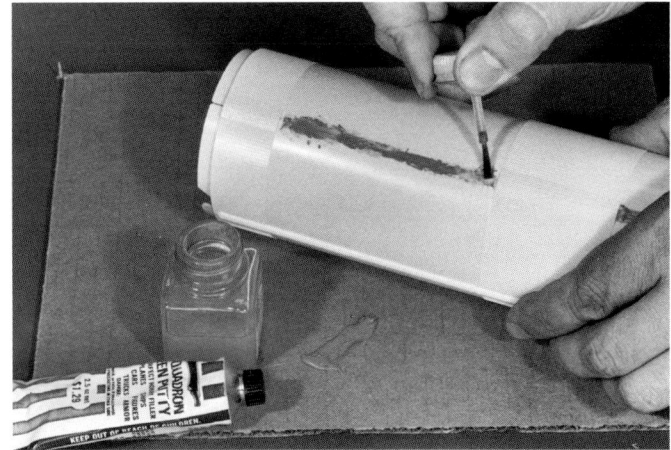
Fig. 15

of the seam was sanded smooth, I still had to repeat the application at a few points. Finally the seams were smoothed to the point where they hopefully would not be noticeable through a coat of paint.

I covered the yellow domes and silver cones with paper toweling (newspaper also will do) and held it in place with masking tape. Then I mounted the stages in turn on a bent wire clothes hanger suspended from the top of my spray booth. After applying a coat of flat white I inspected the puttied areas for imperfections (Fig. 16).

I used gloss white enamel for the final finish coat of paint. I had to apply a little more across the seam area to completely cover the darker putty (Fig. 17).

Once the final white coat had been sprayed on all the stages (as well as the Lunar Module "garage," escape tower, and interstage connecting ring between the first and second stage) and permitted to dry for several days, I began the most meticulous and error-prone part of the project: masking for the second exterior color, black.

I used a sharp new No. 11 blade on my X-acto knife, a straightedge, Scotch Magic Transparent Tape, and a piece of glass on which to cut the tape. I placed a section of tape on the glass, cut a new edge with my knife and straightedge, and began placing the tape along the edge of the paint pattern shown in the instructions and on reference photos. I mounted all the stage assemblies (the twist-ring mounts permit you to assemble and disassemble the rocket for various states of display) through stage two and masked them at one time to make sure the vertical paint separation lines would be in alignment. Later, I cut the tape between stage one and two so the stages could fit into my spray booth and make painting easier in general.

Masking the cone-shaped top of stage two required a little extra thought. Curved or angular pieces are more difficult to mask than flat pieces, and it soon became apparent it would be very difficult to curve the tape to match the horizontal separation line circling the cone about halfway up. So I placed a piece of tape directly over the separation line area. With a small ruler, I measured one-half the distance from the top of the cone and made a tiny pencil mark on the tape (Fig. 18). I moved the ruler to the adjacent rib, measured, and made another mark. After I had completely marked the separation line I made a sharp cut from pencil mark to pencil mark with my knife (Fig. 19). I filled in with more tape below the separation line and also made vertical cuts along the rib lines to complete masking the pattern used on the Saturn Vs. Tape

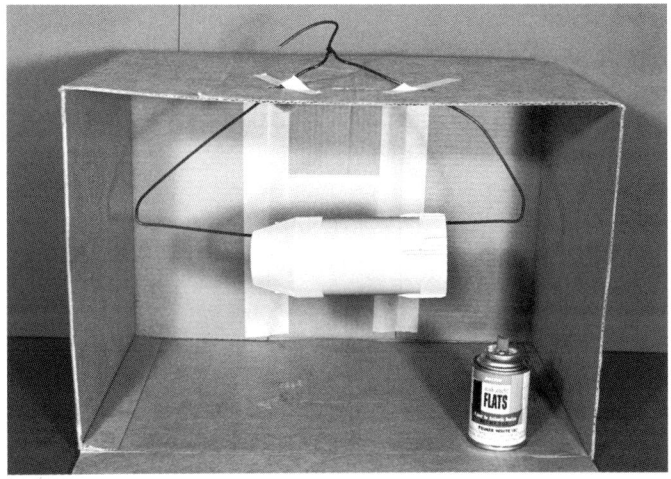
Fig. 16

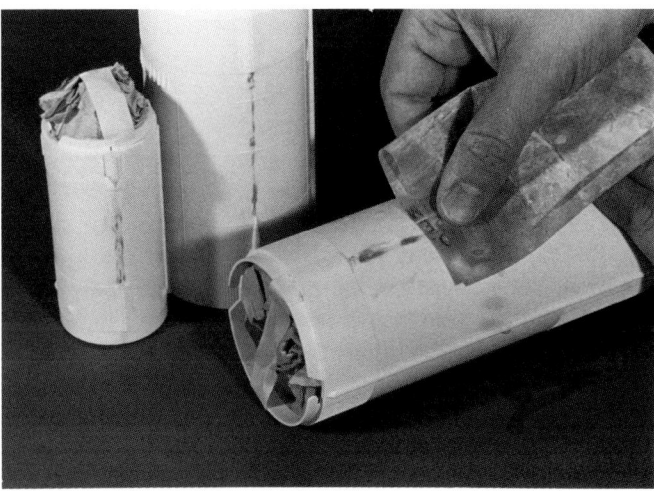
Fig. 17

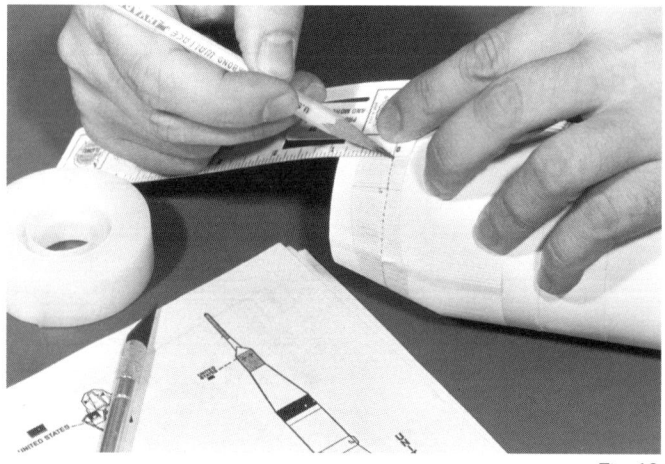
Fig. 18

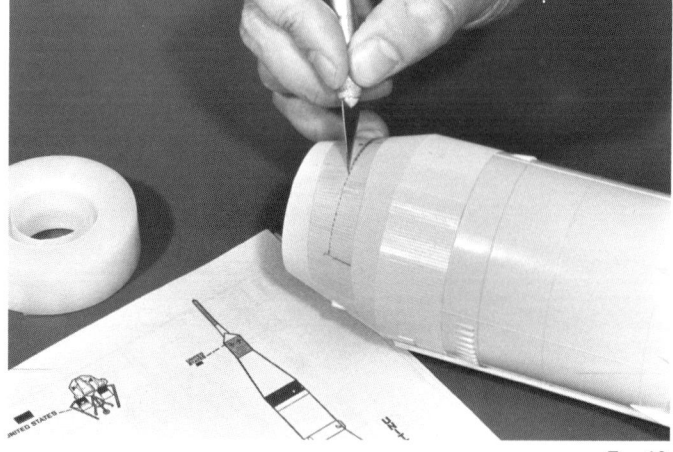
Fig. 19

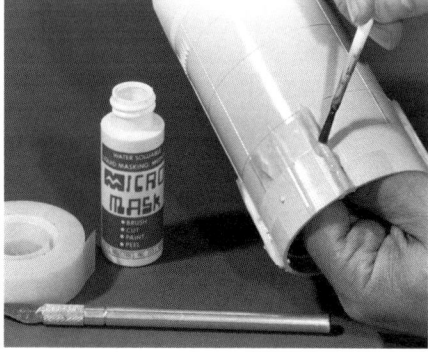
Fig. 20

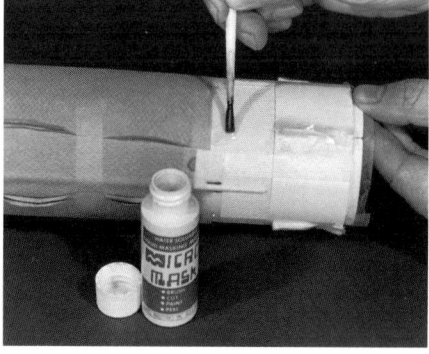
Fig. 21

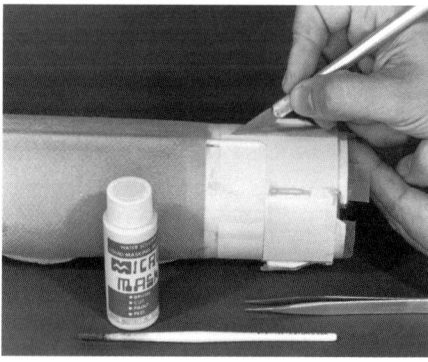
Fig. 22

newspaper or paper towels over large surfaces to protect against overspray.

Ribbed surfaces are also difficult to mask because the tape tends to pull away from the many contours. Thus, paint overspray seeps under the tape and ruins the sharp separation line. Therefore, just before painting, I went over the surface and pressed my fingernail firmly against the tape along the separation lines and along each rib to make the tape adhere tightly.

In difficult masking situations, you can also try a masking fluid such as Micro Mask. I brushed Micro Mask over several of the protruding vanes which would have been awkward and time-consuming to tape (Fig. 20). Micro Mask dries to a rubberlike covering that can be peeled off after painting. It also can be cut with a hobby knife to form separation lines.

I brushed Micro Mask over some of the separation lines where the ribbed surface met the flat surface and the tape was pulling away from the ribs. When dry, I sliced through the material along the separation line and peeled away the part on the area to be painted. The remaining Micro Mask had flowed under the tape and hardened, creating an effective separation line (Figs. 21 and 22).

Masking fluids may not always be for the beginner. If the covering doesn't come off easily, little globs of material may remain and mar the surface as you try to rub and scrape them off. Use a rubber cement pickup, available at art supply stores, to lift remaining material.

As soon as you have finished painting, remove the masking, using a hobby knife to pick at the corner of the tape and tweezers to pull the tape back at an extreme angle, almost parallel to the surface of the model (Figs. 23 and 24).

Of course, you can use your fingers to peel off large pieces of tape, but remember to pull slowly, evenly, and at an angle close to the surface (Fig. 25). In those places where some paint did seep behind the separation line, I used a small-tip brush to carefully touch up or my knife to carefully scrape away. I also masked off the bottom of the first-stage engine cowlings and sprayed them with silver.

At this point, I could have applied decals and a final clear coat to the exterior, but I was anxious to see all the parts assembled. Because I had painted most components before assembly, I still had to attach such items as the stabilizing fins and rocket motors (Fig. 26). Plastic cement will not adhere to paint, so I removed any paint from mating surfaces with my hobby knife and assembled the motors.

I brush-painted the tiny Service Module thrusters yellow while they were still on the sprue. The Apollo capsule's main heat shield was painted gray. Assembly followed.

Glossy paints provide a smooth finish for decals, which I applied in the manner described in detail on page 00. After decaling, I mounted all the stages and gave the entire model a coat of Testor's Glosscote.

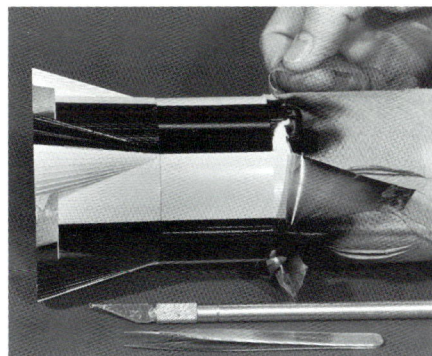

Fig. 23

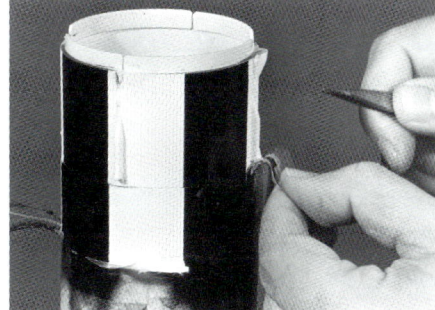

Fig. 24

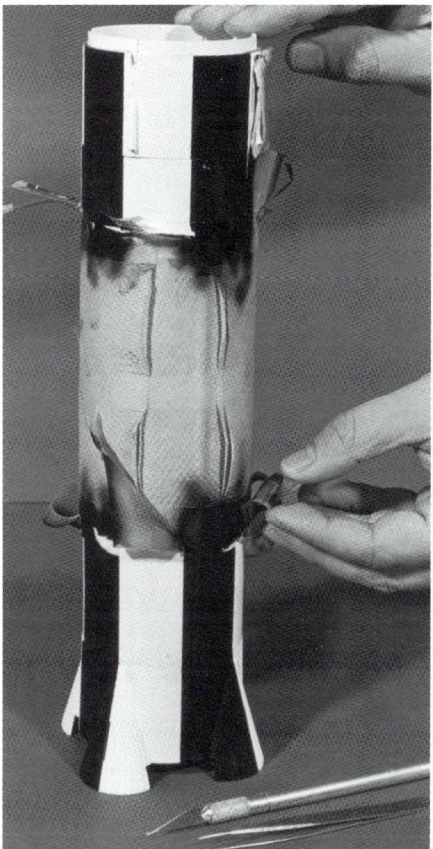

Fig. 25

At last I had a gleaming replica of the rocket that had launched men to the moon. It takes a little time to correctly apply cements and paints, to allow them to dry thoroughly, and to carefully file, sand, and mask. Patience is perhaps the hardest trait to learn, but if you exercise patience with the Saturn V, you'll have a fine display model to show for your work (Fig. 27).

Fig. 26

Fig. 27

BUILD APOLLO BY SUBASSEMBLY

BOB HAYDEN

The basic forms of the Heller Apollo kit are accurate and well detailed. With some minor reworking, proper painting, and a revised, subassembly-oriented construction sequence you can turn out an eye-catching miniature replica of the vehicles used in man's most daring extraterrestrial adventure.

To start, research correct colors and patterns. You'll find the patterns on the craft differed significantly from mission to mission, so you'll have to choose one scheme and stick with it. I found the best source for this kind of information was my public library, in books on space exploration, in *National Geographic* magazine, and in an encyclopedia. While you research, decide whether your model will be finished "new" or "used"; this will make a difference in assembly.

Size up the kit. Study the instructions and look over the parts before you detach them from their runners. The instructions are in French, with English and German translations on the back of the sheet. Write notes on the instruction sheet as you examine and develop an understanding of the model (Fig. 1). Even if you don't intend to follow the sequence presented in the instructions, talk yourself through it, finding each part as it is called for, and noting how and where it fits.

The sequence of painting is the controlling factor in this construction project. My plan of attack was to assemble the major body portions of the model as subassemblies, do any sanding, filing, adzing, and joint filling required, and then prime and spray-paint these subassemblies and the remaining parts. The Heller kit is ideal for a subassembly approach, because it is divided into four portions as part of its design: Command Module, Service Module, Lunar Module ascent stage, and Lunar Module descent stage. I began by making the following subassemblies: Lunar Module cabin with rear RCS thruster mounts (parts 1, 2, 4 and 5), Service Module body (6 and 7), and Service Module engine nozzle (10 and 11) (Fig. 2).

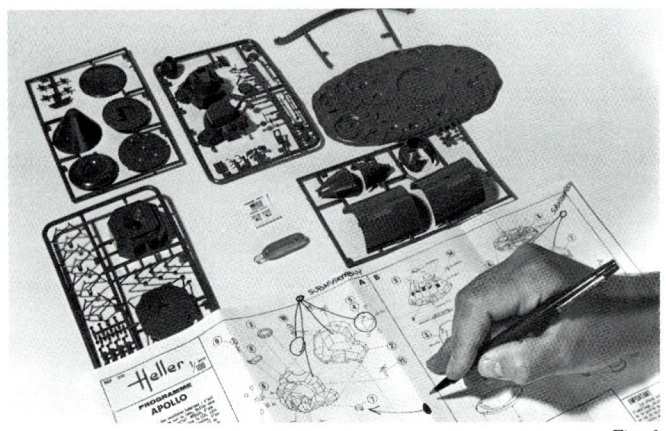

Fig. 1

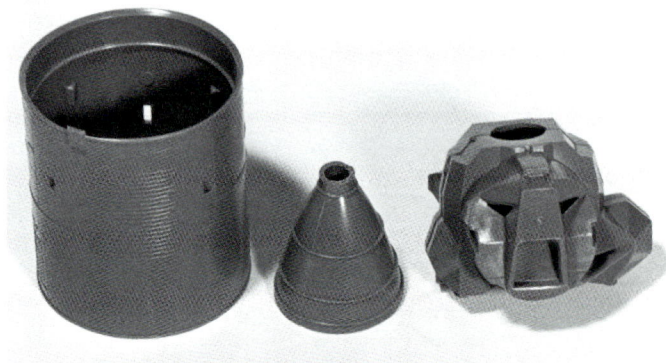

Fig. 2

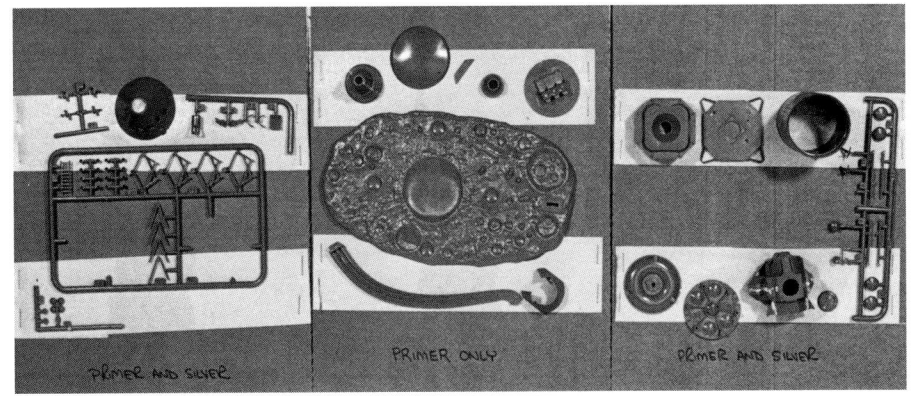

Fig. 3

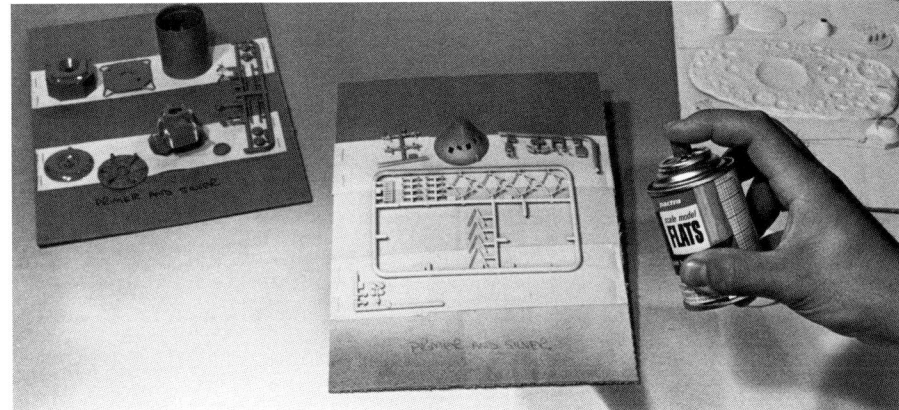

Fig. 4

Little filler putty work is required on the Service Module body halves or engine nozzle, but the Lunar Module cabin halves fit together with unacceptable mismatches. Fill these with putty, and sand with fine and extra-fine sandpaper. Inspect your work, and repeat the filling and sanding process until you bring the surfaces to the proper contour and finish.

Separate all but the smallest parts from the runners and use a sharp modeling knife, sanding block, and files to clean up the mold parting lines and sprue attachment points. Wash the parts, including the subassemblies, in detergent, rinse well with clear water, and let them air-dry. For priming, arrange the cleaned parts on sticky-side-up masking tape stapled to scrap cardboard (Fig. 3). Handle the cleaned parts with tweezers or disposable plastic gloves to prevent contaminating them with finger oils.

Simulating metal finishes effectively is important in any kind of aircraft modeling, and particularly so in space modeling, where many parts on the prototypes are left unpainted as a means of reducing vehicle weight. Before painting or priming the Apollo parts, you'll have to decide how to simulate the polished alloy surfaces of the CM, SM, and LM.

There are several ways to represent shiny metal surfaces; each method has its own advantages and drawbacks. Most common is the use of metallic paint, either enamel or lacquer, that carries millions of tiny metal particles in a standard paint vehicle. Gloss metallics give the better simulation of the shiny surface of the real thing, but hide detail. Flat metallics show up detail well but lack snap and luster.

I chose a semigloss lacquer for my model—a compromise. Whatever type of finish you choose, it is almost mandatory that you apply metallic paint by spraying in order to achieve a smooth, uniform, metal-looking surface. If you use an airbrush and can vary the air pressure, remember that metallic pigments—actually bits of metal—are heavier than regular colors and require increased air pressure (on the order of 25–30 pounds) for proper application.

Another way to simulate metal is by the application of thin metal foil. Applied with an adhesive, the foil burnishes down over panel and rivet detail to form a very effective "skin." The complex, blocky shapes of the LM and the cone form of the CM preclude foil covering for our small-scale model, but foil covering is worth considering for larger-scale efforts.

Rub 'n Buff is a wax emulsion that holds metallic particles in suspension. A paste, it is spread over the model and buffed to a deep luster with a soft cloth. It "shines up" in proportion to the smoothness of the surface on which it is applied, so you must plan on applying either flat or gloss undercoat before applying Rub 'n Buff. The metal simulation is convincing—the disadvantages are time-consuming application and a tendency for the finish to tarnish as the model ages.

Whatever you choose for metallic finish, you'll need to prime or undercoat the parts. You can apply enamel or other plastic-compatible primers with a brush, but the best method is to spray. Use flat white or light gray primer (Fig. 4) to show up imperfections, and sand the trouble spots before final painting. Lacking a standard primer, use flat white or gray paint in the same role.

After checking for parts that must be filled or resanded, respray the parts that will represent metal with your choice of metallic finish. Let this coat cure for at least two days. Next, carefully paint cast-in details and small parts with a No. 00 spotting brush (Fig. 5). You'll find the masking tape flats are convenient "handles" for larger pieces, as are sections of runner left attached to the smaller parts. Handle parts as little as possible from the time you clean them for painting until they are added to the model.

Fig. 5

Fig. 6

The most striking feature of the LM descent stage is the copper-orange Mylar covering. No material is furnished for this covering in the Heller kit, but I felt it is such a characteristic part of the prototype that it had to be simulated. I used copper-colored, lacquered, paper-backed aluminum foil wrapping. This is similar to giftwrap foil, and can be found at art, craft, and stationery stores. Cut a 9" strip of the foil, 15/16" wide, and draw a pencil line 5/16" from each edge. Hold a thin straightedge along each pencil line and bend the foil against it and back upon itself. This will give you a strip of foil 11/16" wide, and the folds ensure that the white backing paper will not show at the foil edges.

Next, gently wrinkle the foil strip. Don't get carried away here; you want the "Mylar" to look creased, not battered! Flatten the foil and apply water-soluble contact adhesive such as Micro Scale Metal Foil Adhesive or Elmer's Acrylic Latex Contact Cement to the back of the strip and the sides of the descent stage body. Let this coat dry for about 40 minutes and apply a second coat. When this is clear, carefully align the foil strip on the body and press it into place, one side at a time (Fig. 6). Overlap the strip at the end, and plan ahead so the overlap comes at the rear of your model rather than beside the LM "front door" ladder as it did on mine! When you assemble the descent stage, distort the foil slightly at the four corners to clear the upper and lower landing gear support arms.

I used liquid plastic cement for most of final assembly, but substituted ACC ("super glue") for joints between prepainted parts where neatness was the most important consideration. ACC is especially helpful in assembling the landing-gear legs. It is ultrafast-setting, and very strong, even where there is little surface area in a joint.

Finishing touches bring any model to life; here are the extra things I added to complete the Heller Apollo kit. First, I finished the moon landscape base with dead-flat gray and gray-brown colors. The flat paint is a pleasing contrast to the semigloss metallic silver and shiny copper foil areas on the spacecraft. Next, I made a flag similar to those planted by the Apollo astronauts by applying flag decals from a model railroad "Bicentennial" locomotive decal set to thin styrene sheet. I then cemented the flag to a length of stiff brass wire placed in a hole drilled in the base.

I did not weather my model, although it has good possibilities for a weathering job. The heat shield on the CM can be painted to show the scorching that takes place during re-entry, and the LM landing gear foot pads can be dusted with gray moon dust made by powdering gray and tan pastel chalk sticks.

The final extra detail I added was an astronaut figure. This came from an out-of-production Revell 1/96-scale Apollo kit, and I consider it important to the overall display because it lends an easily understood scale element.

THE ULTIMATE STARSHIP

RICHARD G. VAN TREUREN

Alternate markings: More accurate markings are available from Estes Industries Penrose, CO 81240. Ask for U.S.S. Enterprise decal, part No. PN 37078. Enclose $1.75, which includes postage and handling.

Although several different models have been offered over the years, we will limit our discussion here to making accurate models of the pilot-film and production versions of the *Enterprise* from the AMT kit. Overall, this kit is an excellent miniature, and in one sense, it is absolutely accurate because several of the kits were assembled, damaged, and used in second-season episodes.

Unless you have an older (large box) AMT kit, you'll need an Estes *Enterprise* rocket kit for the decals because the more recent (small box) AMT kit includes inaccurate markings.

Before you start on the model, examine your reference material. Disregard photos of the smaller yard-long model (usually identifiable by the sky backdrop). Use photos of the 11-foot model as your guide, including those presented here, and ignore all but the corrected drawings that appear in this book. I've found that the best available references on the *Enterprise* are film clips of the model in the studio and publicity pictures that were taken just after the model was modified in 1966.

Some small inaccuracies in the AMT kit are virtually impossible to correct—the slight taper of the nacelles is not duplicated in the kit, for instance, and it would be very difficult to add. Depending on which version of the *Enterprise* you choose—pilot-film or production—there are several things that will greatly add to the accuracy of your AMT-based miniature.

Regardless of the version you choose, start by removing the raised grid lines on the top surface of the primary hull. The ribs are interesting detail, but the *Star Trek* original was a smooth spacecraft. The key to the job—which will take an hour or so—is a sharp hobby-knife blade. Work slowly and use a No. 10 X-acto blade to cut, peel, and scrape away the ribs. It's better to leave a little of the rib material to be sanded away than to dig the knife blade into the surface.

Sand the upper surface of the primary hull to remove the last traces of the grid. This is a small-scale model, so the finish must be nearly perfect to appear realistic. Start by wet-sanding with 400-grit wet-or-dry paper, and then use 600-grit paper, also wet. Inspect the surface frequently, using strong light from the side, and keep at it—this kind of work is mostly elbow grease. When you are satisfied with the finish, prime the model, let dry, and dry-sand with Flex-I-Grit 8 micron silicon carbide sanding film. This should give you a smooth surface for the final silver-gray paint coat.

Either version also requires filling the three cup-like depressions on the underside of the primary hull. Add aircraft-type navigation lights next: red on the port side, green on the starboard.

The most recent pressings of the model have omitted the navigational beacon at the top rear of the secondary hull. Any bubble-shaped item could be attached to form this beacon. You can make such a bubble by heating the tip of large-diameter sprue with matches until it melts to the proper shape. Saw the stem off flat, and cement the bubble in place.

Modeling the pilot film version. If you decide to model the pilot-film starship, remove the cross-hatched, waffle-textured panels on the inner sides of the nacelle support pylons. Likewise, sand away the screening texture found in the inboard slots of the nacelles.

The main sensor screen on the pilot-film *Enterprise* was considerably larger than that on the AMT model. To enlarge it, add to the outer edge a ¼"-wide gasket or a sheet-styrene ring, fill in the gaps with putty, and smooth to the proper contour. Trim the stem or enlarge the mounting hole so the screen can be drawn closer to the secondary hull.

Since the hangar bay was not a part of the ship until regular show production, remove the observation post atop the hangar-bay doors. The bridge section of the primary hull must be raised. A common fuse-holder cap, available at any electronics parts store, fits well and can be sanded to the proper shape.

Carefully locate the center of each forward nacelle cap, drill, and add a small spike. An electronics store can provide electronic module gold turret terminals that will serve nicely here. Also, remove the globe-shaped portions of the rear nacelle end caps, and flatten those areas.

To finish the pilot-film version, use a satin, semigloss coating—not gloss, not flat. To be strictly accurate, the model should not be weathered. Decals from the Estes rocket kit can be trimmed and added to simulate many details on the pilot-film model. Don't forget to put the numbers for the underside of the hull on backwards.

The production *Enterprise*. Start on a production version of the starship by reducing the diameter of the main sensor screen about ⅛". As on the pilot-film model, the screen must be mounted close to the hull. Add raised dots made from pinheads to the flat forward ends of the inner intercoolers. If you have an older large-box kit, the globe-shaped nacelle end caps can be simulated with small marbles or large ball bearings.

The lip of the bridge cap must be filled in to enlarge it, but be careful not to increase the height of the bridge in doing so. After adding the navigation lights mentioned earlier, position smaller lights in line with them on the underside edge of the primary hull. As before, fill in the three cuplike depressions on the saucer underside.

Since the production *Enterprise* was subtly weathered, the model should be finished in a flat silver-gray. Here again, details from the Estes rocket decal sheet can be added to restore any raised details sanded away in the crack-filling and modification work. Finally, the numbers on the underside of the primary hull can be installed either forward or backward, since they were seen both ways during the run of the television program.

KIT CONVERSION: ENTERPRISE TO TUG

DON KLEIN

Although the various styles and classes of United Federation spaceships derived from the *Enterprise* design are a creation of those who established a post-*Star Trek* mythology, the designs nevertheless can be intriguing to modelers wishing to engage in some simple kit conversions.

Publications such as the *Star Fleet Technical Manual* (Ballantine Books, 1975) show drawings for Destroyer, Scout Dreadnought, and Transporter-Tug ships, in addition to the Heavy Cruiser (*Enterprise*) configuration.

Any of these derivative designs can be built using the parts from the AMT *Enterprise* kit. I chose to build the Transporter Tug, which meant that I would have to build a transport container in addition to the parts available in the kit.

For the transport container, I was able to obtain a length (7") of 2⅜" acrylic tube. A length of 2½" plastic plumbing pipe should also work. Some of these items may be hard to locate, but you are free to choose your own materials (including household containers) that form a suitable cylinder. A big box of leftover parts from other plastic kits is almost an essential item if you plan

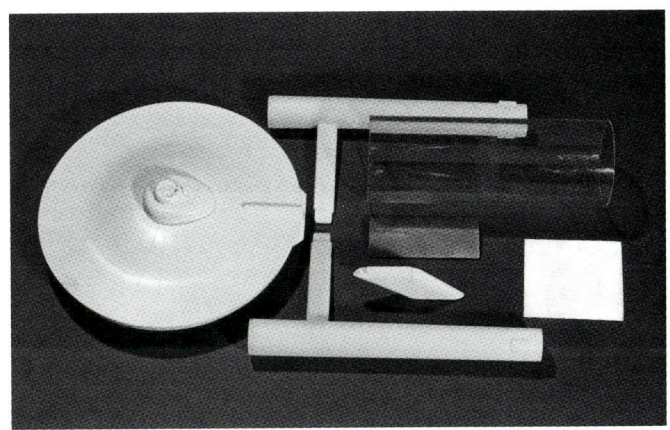

Major components used in making the Tug were the saucer, nacelles, and central strut (sawed from the secondary hull) of the AMT kit, plus bomb-bay doors from a Monogram B-29, a length of acrylic tube, and sheet styrene for the tube ends.

to do many kit conversions, and my leftovers included bomb-bay doors from a 1/48-scale B-29. These would form the towing pad.

I spread the basic components out on the table to get a better idea of what was needed and how everything would go together.

I began my work with the primary hull. The major change was that I discarded the bottom dome and added a support for the main sensor. In place of the dome, I cemented a thin plastic disc, then extended a 20 mm gun barrel (almost any polelike piece will do) from Monogram's Panzerspähwagen eight-wheeled armored car. Using ACC adhesive, I placed a BB at the end of the gun barrel and then attached the sensor dish (after cutting away a portion of its attaching peg) to the BB, facing it forward.

The warp-drive propulsion units required a little more work. Because the units are supported by downward projecting pylons on the tug, I was able to simply reverse the positions—the kit's right propulsion unit became the left and the left became the right. The intercoolers, however, still had to be moved to the top of the warp drives. The unneeded location holes were filled with sprue and sanded over. The attaching pegs on the intercoolers were removed, and the units were placed flush on top of the warp drives and cemented with Micro Weld.

The warp-drive pylons were attached directly to the bottom of the primary hull. To get the proper placement angle, I used a razor saw to cut off the tip of the pylons at a 45-degree angle. A piece of 1/8" sheet styrene was cemented to the end of each pylon, allowed to dry, then filed and sanded to shape. Later the pylons were cemented in place.

I placed the bomb-bay doors that I used as the tow pad over the acrylic tube and cemented them so they would conform to the cylindrical shape.

I used .020" sheet plastic for the ends of the transport container. I placed the sheet plastic under a 3" circle template and made a continuous cut with a heavy straight needle (not a straight pin) until the plastic was cut through. I made covers for each end, attaching them in place with ACC. The plastic cements would not have bonded to the acrylic tube securely enough to endure filing and sanding. After the sheet plastic was firmly secured, it was cut and filed to the contour of the tube.

The strut joining the primary hull to the towing pad was the same strut furnished in the kit except that it had to be severed at the bottom where it was molded to the *Enterprise*'s secondary hull.

I airbrushed the Transporter-Tug with a mixture of Pactra primer white and Pactra aero blue. I also gave it a coat of Micro Gloss heavily diluted with Micro Thinner, resulting in an even matte finish. I used several decal sheets to come up with enough letters to form "Star Fleet Transport Command" on the sides of the container.

DECALING — A MOST REWARDING ENTERPRISE

DON KLEIN

The AMT *Enterprise* assembles easily, with only a modest amount of putty work. Decaling, however, is such an important part of building this model that the technique described here applies to all models.

After seam filling and sanding, which had to be done with extra care because of the extreme brittleness of the plastic in this kit, I replaced the sanded-away windows with .040 sheet styrene cut into 1/16" by 1/8" rectangles.

Mine is a TV watcher's model, so I airbrushed the *Enterprise* Humbrol underside white, since that's the way it appears on TV. I then sprayed Floquil gunmetal over the outside of the nacelle end caps after the balls had been masked to protect the underside white.

The warp drive domes at the front of the nacelles were sprayed Pactra brick red (Fig. 1). The main sensor dish was painted copper; the back side of the dish and

Fig. 1

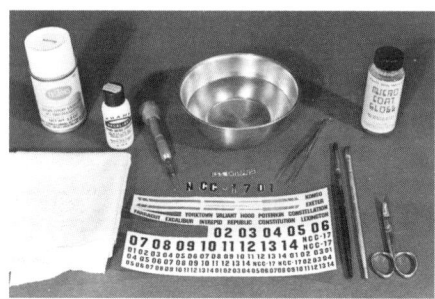

Fig. 2

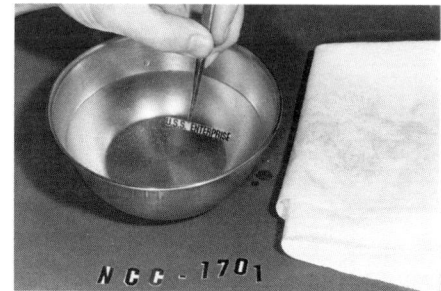

Fig. 3

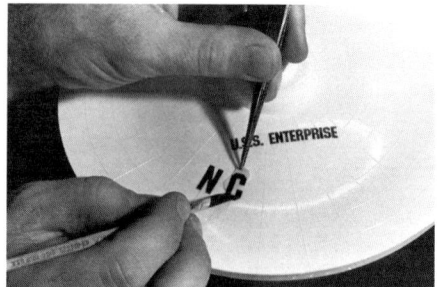

Fig. 4

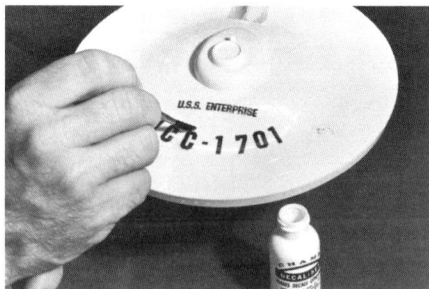

Fig. 5

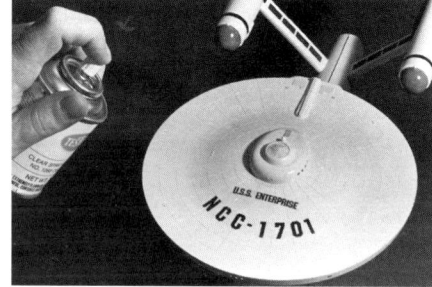

Fig. 6

front of the secondary fuselage were painted flat black. I used a No. 000 brush to apply Pactra tinting black to the raised sheet-styrene windows. Tinting black was also used on the strut, nacelle, and intercooler screens. The model was now ready for decaling.

The surface of the model must be clean, dry, and free of oils, fingerprints, or other contaminants. Use small pointed scissors (embroidery or nail scissors) to trim the clear decal film close to the printed lettering and designs (Fig. 2). The main object is to de-emphasize and hide this film, so trim it back as much as possible.

The success you have with the decals can also depend on the paint you are using. Decals adhere far better to a smooth finish (gloss paints) than they do to a rough finish (flat paints). If you have painted a model with flats, you might want to apply a coat of a clear gloss over all of the model. I gave the *Enterprise,* which I painted in flats, a heavy coat of Micro Gloss prior to decaling.

Place the decal pieces, one at a time, in a shallow dish of lukewarm water to soak the film from the backing sheet (Fig. 3). The decal is ready to be applied when it slides freely on the backing sheet, usually after 10 or 20 seconds of soaking. Use tweezers to place the decal and backing paper on a clean paper towel to drain off excess moisture. Position the decal on the model and gently slide the backing paper out from under it (Fig. 4). Carefully adjust the position of the decal by gentle prodding with the tip of a watercolor brush, and blot up all remaining water with the corner of a paper towel. Gently press the decal against the model surface with your fingertips, and try to work air bubbles out from the center of the film to the edges.

Next, apply decal-setting solution (Fig. 5). There are several brands, but all are solvents that soften the decal carrier film and allow it to conform to the model. Be sure to use the brand recommended by the manufacturer. Brush on the first application of setting solution, and leave the decal alone while it works. Because of the raised panel lines on the *Enterprise,* I even brushed a drop of setting solution on the surface of the painted plastic before I laid down the decal.

When the decal is dry, use a needle to prick any bubbles in the film, then brush on more setting agent. Repeat this process until the decal hugs the surface perfectly.

Give a model at least overnight to dry, then apply a clear flat coating to further hide the edges of the decal film and protect the lettering during handling (Fig. 6). Any of the commercial clear flat products are suitable, including Floquil flat finish, Pactra clear flat, Scalecoat flat glaze, and Testor's Dullcote. Several of these coatings are available in spray cans.

AMT's Klingon Battle Cruiser is a formidable-looking craft, made more realistic with miniature bulbs and fiber-optic strands. FSM photos by Chris Becker.

LIGHTING AMT'S KLINGON BATTLE CRUISER

Smaller bulbs and fiber optics light up the darkness of space

MARK P. WILSON

When the Klingon Battle Cruiser first decloaked on national television in *Star Trek: The Next Generation,* fans thought this evolution of the old series D-7 cruisers was far out.

As radical to its predecessor as the 1701-D *Enterprise* was to the earlier Federation starships, this design is clean, compact, and powerful. There is no room for families or scientific exploration—only weaponry.

The entire pincer-shaped forward section is a huge disrupter pod, the Klingon equivalent of a phaser (except it hurts more). Though physically smaller than the Romulan War Bird or Federation *Galaxy*-class vessels, this ship would be on equal footing in a tangle with either.

I decided to try lighting AMT/Ertl's kit when I saw the clear red engine pieces and the cleanly marked ports. Having used small incandescent bulbs to light other kits, and faced with serious internal space constraints, I settled on a combination of Mini-Maglite bulbs and a halogen flashlight bulb coupled with fiber optics to provide the lights. Using flashlight bulbs also greatly simplified the power

Fig. 1. Mark painted the interior chrome silver to prevent light from showing through the hull.

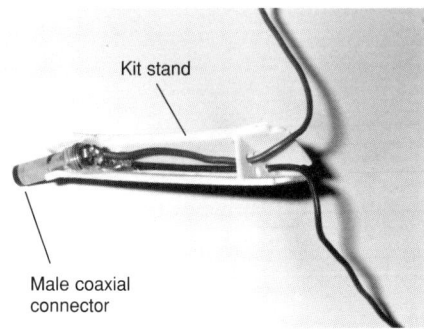

Fig. 2. The battery pack for the lights is wired to the model through the stand and this coaxial connector.

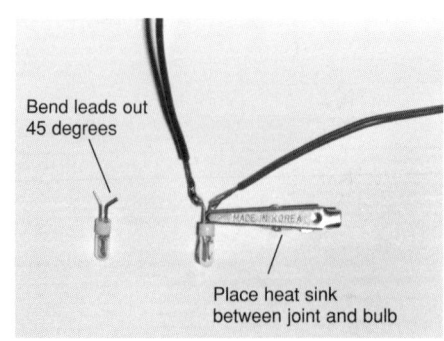

Fig. 3. Socketless mini-bulbs were connected directly to the wires because of confined space in the model.

requirements: 3 volts, from a battery or transformer.

You need a plan. If you're going to light a model such as this one, spend several hours drawing the hull layout and planning bulb locations and wire routes. This ship has three blinking navigation beacons as well as steady engine and port lights.

Since I planned to wire the bulbs in parallel, I had to run the blinking bulbs on a separate circuit to keep them from dimming the other bulbs when they blinked. I made a rough layout of the fiber optics and electrical wire routes to avoid conflicts.

Here is a list of the tools and materials you'll need for this lighting project:
- Pin vise
- .025" and .035" drill bits (No. 72 and 65)
- Gap-filling super glue
- Super glue accelerator (or super glue accelerated and filled with baking soda)
- .020" and .030" fiber optics (found in craft and hobby shops)
- Hair dryer
- Soldering iron
- Small heat sinks
- Epoxy putty
- Wire working tools
- Power supply (either a small train transformer or two D-cell batteries)
- Flashlight bulbs
- 18-gauge multistrand wire

Building to plan. Starting with the areas where bulbs and wires are mounted, build up the lower hull, side walls, and the bottom of the engine pods into one subassembly. Next build the top and side walls of the weapon pod.

Spray the interior of these subassemblies bright metallic chrome to make them opaque and reflective, Fig. 1. You are now ready to add lights.

A drilling fool. The lighting is done in three steps: drilling portholes, wiring bulbs, and fitting the fragile fiber optics. Drilling the portholes helps you figure out where the wires and fiber optics will be placed. Drilling is tedious, but I recommend using a pin vise instead of a motor tool to keep tolerances tight and rows even. Drill all the holes in one sitting, starting in the least obvious spot; you'll get better as you go along.

The ports on this kit are indicated by small raised squares. Leave the raised detail to help set the optics and sand them flush later.

The wiring. I modified the kit stand to route wires from the power source into the model. The connection is a 5 mm-diameter coaxial female plug inside the kit and the male jack extending from the stand, Fig. 2.

Because there is little room in the engine nacelles, solder the mini-bulb connections to wire instead of using a socket. Wire the mini-bulbs and halogen bulb (in a socket) in parallel.

Soldering wires to bulbs is tricky and you could blow bulbs apart if you're not careful. Wear eye protection and a dust mask. Bend the bulb leads out 45 degrees with needle-nosed pliers, then apply a heat sink (a small alligator clip will do) between the solder joint and the bulb, Fig. 3—this will help protect the bulb from the heat of the soldering iron.

Strip the insulation from the multistrand wire back ¼" and wrap the wire and lead together. Apply the iron to the wire. When the wire insulation starts to melt, touch the joint with solder and let it flow into the connection.

Repeat the process for the blinking light set and the bulb in the disrupter pod. Test all lights with the power source to make sure the connections are good. Now bend the wires to conform to the interior of the model, abut don't bend at the

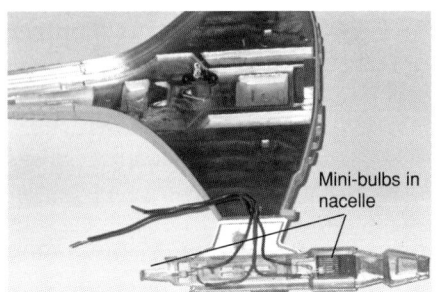

Fig. 4. Wires from the Mini-Magilight bulbs in the nacelles are routed through the hull to the stand mount.

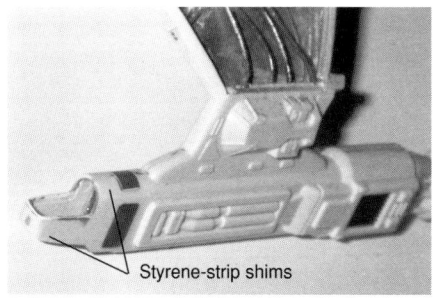

Fig. 5. Mark improved the fit of the clear-red parts on the engine nacelles with styrene-strip shims.

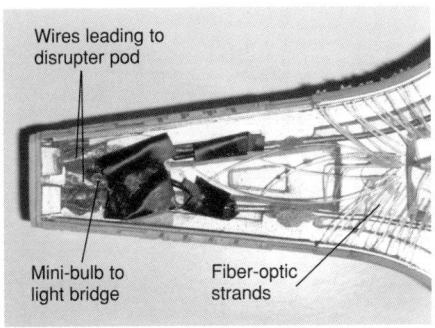

Fig. 6. This view of the forward hull shows the wires leading to the disrupter pod, the mini-bulb that lights the bridge, and fiber-optic strands already installed along the hull edges.

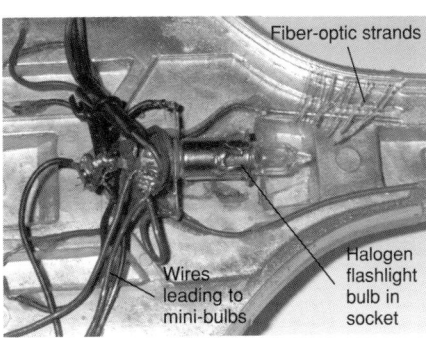

Fig. 7. Fiber-optic strands point toward the main halogen flashlight bulb.

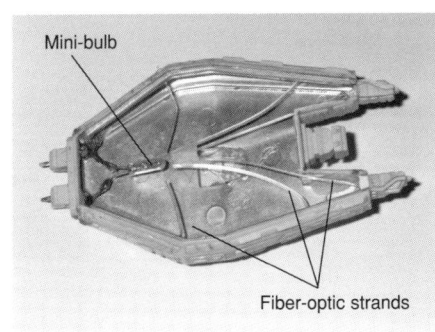

Fig. 8. The mini-bulb in the disrupter pod lights fiber-optic strands running to ports on the edges.

soldered joints. When the bulbs and wires are in place, anchor them with super glue, Fig. 4. The bulbs in the warp nacelles give bright light through the clear plastic, but they're back far enough to remain hidden. Sanding the inside of the clear parts also helps diffuse light.

I modified the clear red nacelle panels to fit flush with the surrounding structure by using styrene-strip shims, Fig. 5. If you need to add clear red material around the kit panels, mix Tamiya clear red paint with Micro Kristal Kleer and brush it into the gaps.

The impulse engine bulbs were positioned to also light rear hull fiber optics. Mount the last main hull mini-bulb directly below the bridge, and point it upward, Fig. 6.

To provide light for the central hull fiber optics, mount a halogen bulb socket to the female coax jack plug with a blob of epoxy putty. Figure 7 shows how congested the interior of the main hull is with some of the fiber optics and the wiring for eight mini-bulbs in place. Mount a mini-bulb in the disrupter pod, Fig. 8.

Wire the blinking lights into a separate circuit. I used a mini-bulb wired in series with a blinking Christmas-tree bulb mounted inside the model stand. Wired this way, the holiday flasher causes the other bulbs wired to it to blink as well. Build a light-tight box around each mini-bulb to prevent its light from being picked up by the fiber optics.

Fiber optics. Installing the fiber optics is the most challenging task. Fiber optics allow you to produce tiny pinpoints of bright light all over the model, representing ports of lighted compartments. The light source for most of the fiber optics is the halogen flashlight bulb in the center of the ship. The strands transmit the light from one end close to the bulb to the other end placed at the drilled-out ports.

As small as they are, fiber optics can eat up space as you bundle them in crowded assemblies. Fiber optics can be bent and formed to some degree, but if you bend too much, the fiber cracks and will not transmit the light.

Route the strands as directly as possible and place them perpendicular to the surface at the exit point for best light transmission. You may have to layer rows of fibers over each other, and you can glue the layers together with super glue. Watch out, though, some super glues may melt fiber optics.

When cutting or bending fiber optics, heat them first with a hair dryer set on high. Clip them about ½" longer than you need with a sharp parts cutter or nail clipper. Heating will keep the fiber soft as you cut, preventing the ends from cracking or fraying. Keep the ends of the fibers about ¼" away from the light source to prevent melting or scorching.

With a strand in place, use a small drop of super glue on the

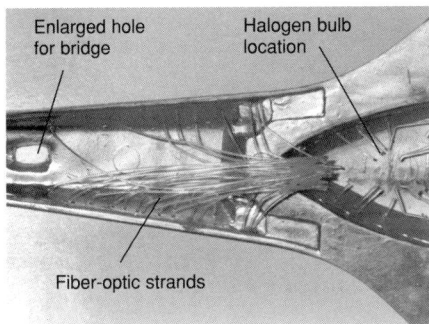
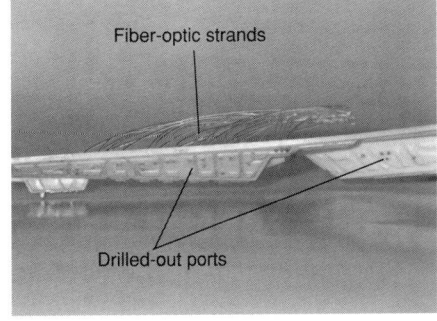

Figs. 9 and 10. Fiber-optic strands are bundled and glued together. They point inward toward the light source.

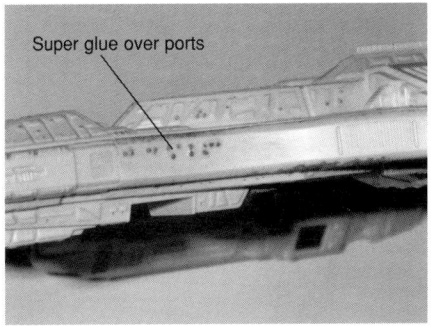

Fig. 11. The ends of the fiber-optic strands meet drilled-out ports along the hull edges. Super glue that fastens the strands can be sanded flush to the surface of the hull.

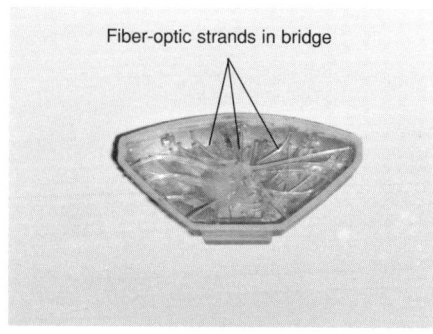

Fig. 12. The bridge also requires a bundle of fiber-optic strands.

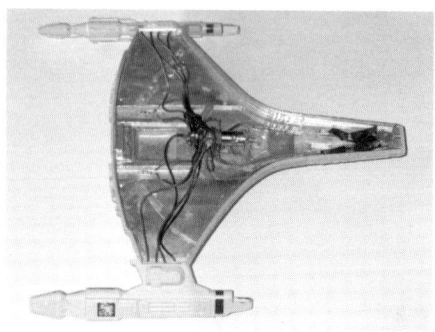

Fig. 13. Minus the disrupter pod and upper hull, the model shows myriad wires, bulbs, and fiber-optic strands.

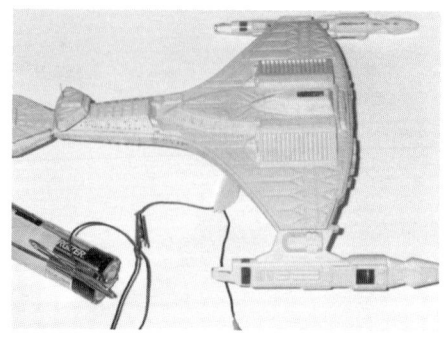

Fig. 14. Mark occasionally checks the lights by connecting the battery pack.

outside surface of the port, pull the fiber back slightly into the hull to get a good bond, and then apply accelerator or dust with baking soda.

Figures 9 and 10 show the fibers installed in the hull. Note how dozens of fibers are glued together facing the bulb location. Sand off the glued ends at the surface of the model, Fig. 11.

Repeat the process for the bridge, Fig. 12. I enlarged the opening at the bridge mount to make it easier to route the fibers. This weakened the bridge mount, and required lots of super glue, but the glue and fibers created a sturdy stem to support the bridge.

Figure 13 shows the completed lower hull half without the disrupter. Final assembly involves stringing fibers through ports in the upper half of the hull and gluing the hull together. Test the lights once again before painting, Fig. 14.

Painting. How do you mask a bazillion (OK, maybe less than 100) .020" ports? Use a sharp round toothpick to apply a raised blob of white glue to each port. Mask the large lighted areas with masking tape.

I mixed eight parts of Testor's Model Master pale green (FS 34227) to three parts white and one part blue for the exterior color. Overall, AMT's paint guide is good, but the ports should be clear, not red, and the rear disrupter banks should be flat orange rather than bright yellow.

After each coat, turn on the lights to make sure the paint is opaque enough to hide the lights inside. Once the paint has set, use a sharp blade to prick the white glue blobs off the ports, then touch up if necessary. Fibers that don't transmit the light adequately can be painted black to simulate unlit compartments.

Voilà! With just a little hard work, persistence, and mindless repetition, you'll have a definite showstopper!

SOURCES

• Fiber optics and lights: Precision Scale Model Engineering, 33 Harding St., Milford, MA 01757, 508-478-3148
• Light bulbs, sockets, wire, and solder: Radio Shack
• Micro Kristal-Kleer: Microscale Industries, 1570 Sunland Lane, Costa Mesa, CA 92626

Photo by Mack Harris

PAINTING AND MOUNTING A KLINGON BIRD OF PREY

RUSTY WHITE

The Bird of Prey is a *B'rel* class scout ship with a complement of about a dozen Klingons, including officers and crew. Small, fast, and very maneuverable, it is equipped with a cloaking device and can easily reach warp speed. Armament includes photon torpedoes and phaser cannons on each wing, making it heavily armed for a ship of its size. It utilizes the advantage of atmospheric travel and landing capability over most Federation starships, using its large wings, which are moveable to attack and cruise positions. The Bird of Prey is a formidable adversary if provoked.

Assembly. AMT/Ertl has hit a home run with this kit. The overall fit is very good. The assembly of the Bird of Prey is straightforward with only one problem: when the wings are assembled in the attack position there is a large gap between the wings and fuselage on the underside, exposing the inside of the model. This meant I had to use strip styrene to build a box hiding the hole. After the model is painted dark green it's unnoticeable.

There was a good-size gap in an unreachable space on the neck between the main fuselage and the bridge. The only way to fill such a gap is to use thinned white glue and brush it on.

Hollowing out the phaser barrels adds realism to the model. To hollow out gun barrels use an X-acto knife with a new No. 11 blade, or with a twist drill if you have one. Use a magnifier to precisely place the drill bit or knife tip dead center on the barrel to be hollowed out. Once it is in place, use light pressure and twist. Be careful not to twist too long, or you will damage the barrel.

Base/stand. After looking over the plastic stand that came with the kit, I decided to make my own. I wanted the model to stand a good 12" above a wood base so the underside would be visible. I also wanted to be able to remove the base and stand from the model, making it easier to transport.

The design called for a stand I have used on other models. It's very strong, looks good, and allows the

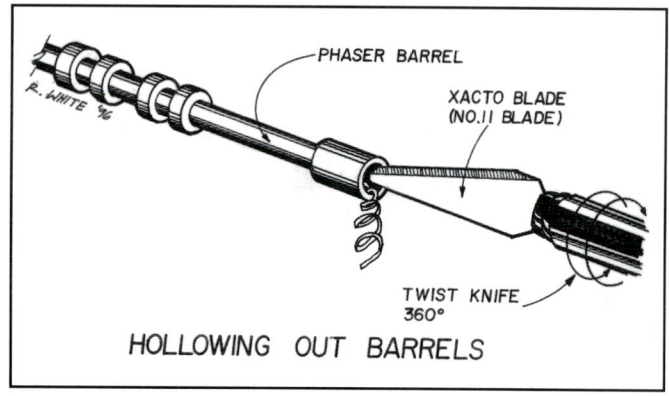

The only equipment needed to hollow out barrels is an X-acto knife.

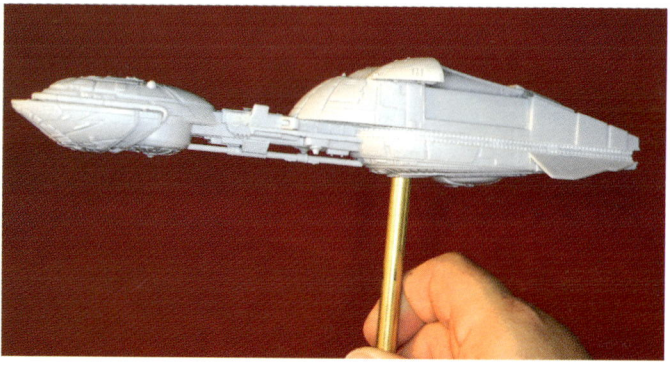

With only the fuselage to work with, mounting the tube in the fuselage is much easier. Don't forget to leave off the warp engine until last!

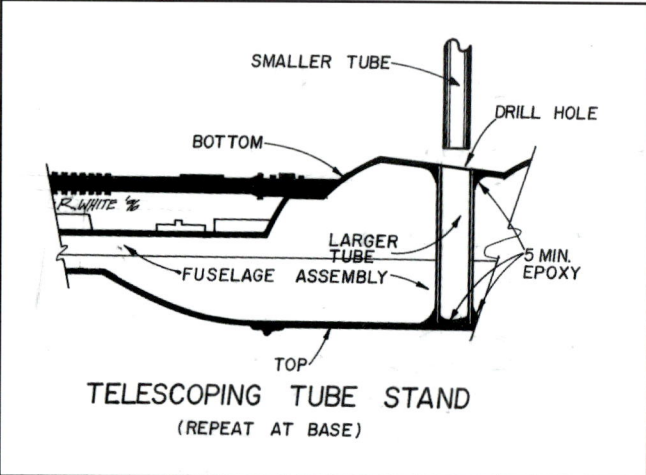

A section cut through the fuselage

The stand with star background base. The model swivels freely and the tube can be removed for shipping.

model to rotate. It is basically a telescoping brass tube with a larger tube inserted in the base and model and a smaller tube between.

Get some brass tubing at your hobby shop. You'll need two sizes, one the next size smaller than the other. The fit should be so exact that when one tube is inserted in the other, they slide easily with no gap between them.

Drill a hole in the base and model while you have only the fuselage assembled, so it is easier to work with. It's important not to attach the warp engine on the back. You will need this hole to apply epoxy glue to the tube once it's in place. Take your time and cut a hole that will result in a tight fit. This is essential when fitting the interior tube.

Cut two lengths of larger tubing—one to fit inside the model and one in the base. Measure the model carefully so you don't have any tubing above the plastic. After cutting the smaller tubing to the desired length, place the section of larger tubing in the base and glue with 5-minute epoxy. Insert the smaller tube while the epoxy is still wet and adjust to a perfect 90 degrees using an angle. Once it's in place, repeat the process on the model.

Painting. Anyone who has built any *Star Trek* models knows there is a lot of painting involved. This one is certainly no exception. As a matter of fact, I spent five times as long painting the Bird of Prey as I did building it. The instructions call for no less than ten colors. I used Testor's Model Master colors throughout, substituting dark green for the Sac bomber green and medium green for the pale green. The one thing I did not want was to have a monolithic green-colored ship. This called for painting some panels colors that differed from the instructions.

First paint the entire model medium green and set it aside to dry overnight. The next day the real work starts. You will paint virtually every panel on the ship one of three colors: dark green, zinc chromate, or medium green. This calls for a lot of brushwork. When using a flat sable brush, mix the colors about 50/50 to prevent brush strokes.

Always start with the lighter colors and work progressively darker. Follow the instructions to the letter when it comes to the zinc chromate panels. With the dark green panels the only place you should deviate is on top of the wings. Virtually all the panels there are supposed to be medium green (the monochromatic effect I feared). Randomly select about half of them to be dark green. Take care to paint the same panels on both wings. Do the rest of the painting according to instructions. After the paint dries overnight, touch up wherever the paint strayed.

The wash. If ever a model cried out for a wash and drybrushing, this does. With its reptilian appearance and heavy raised panels it would benefit greatly from this technique.

A wash is nothing more than thinned-down paint. Once applied the color concentrates in cracks and crevices, thus darkening the low spots. My wash is oil paint thinned with Permtine (a turpentine substitute found in arts and crafts stores with the oil paint). Klingons were known for functional ships, not pretty ones, so you will want a wash that looks somewhat dirty. Buy two colors of oils, burnt sienna and black. Mixed together they make a chocolate brown wash perfect for the dirty appearance you want.

The darkness of the wash is best learned through experience, but a good starting point is that it should be translucent when painted on white paper. If it's too thick, the wash will darken the overall color too much; if it's too thin, you'll have to use multiple coats to achieve the desired appearance. The panel colors should be darkened by the wash, toning them all to a more consistent color. The zinc chromate panels really pop out on the darker green ship, so this will be important.

In order for the wash to flow over the model with no brush strokes apply a wash of clean Permtine to the entire model with a soft wide brush. Allow the wash to dry about 30 minutes. Now with a soft cloth diaper, lightly wipe away some of the wash in the direction the ship would fly through the air. This gives the natural weathered effect of atmospheric flight.

This technique also works like drybrushing—wiping away the wash highlights the raised areas of the model. To tone down the zinc chromate panels, apply another coat of wash over them. Wait another 20 minutes and wipe the panels again. If they are still a little bright, repeat the technique. Let this dry overnight, since you will be handling the model for the wash on the bottom, where you will repeat the same technique. Once it's dry, use a fine-tip brush and apply the wash again to all the depressions between the panels. Allow this to dry one hour before wiping away any wash that has spilled on top of the panels.

Photo by Mack Harris

Close inspection of the wings reveals the weathering from atmospheric flight. Don't overdo it!

Drybrushing is a simple technique that effectively highlights the high points of the model against the darker washed depressions. Most of the world's best modelers use this technique. It is particularly useful to pop out detail in monochromatic color schemes.

You will need a wide, flat short-bristled brush, because you'll have to apply pressure on the brush to smooth out some areas while you're drybrushing. Normally, you'll use the same colors as before, only slightly lightened. Since the wash has darkened the finish, use the same colors as before on the zinc chromate panels. To blend the colors a little, drybrush the entire model with medium green.

Dip the flat brush in the paint and brush the paint onto a piece of paper until there is almost no color left in the brush. Start brushing gently, using light pressure until the high spots begin to pick up the color from the brush. Work slowly and don't be in a hurry. As you use up the paint in the brush, start applying more pressure to highlight all the raised areas. This scrubbing action

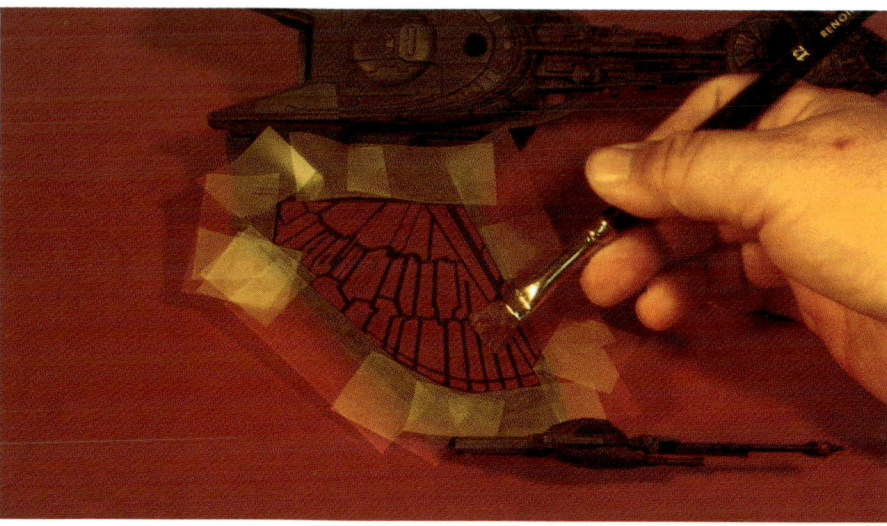

(Left) To blend the panels closer in color I drybrushed the entire model medium green. Note how all detail pops out, even in the darker areas. (Above) I had to tape off the red parts to keep red from the surrounding parts. Note the wide, flat brush.

Markings were easy to paint on the model. There were only four in the kit. Photo by Mack Harris.

I painted the windows camouflage gray, as if lighted from inside.

blends the paint evenly. You will need a smaller brush for tight areas.

I would suggest you practice on an old model until you get the hang of it. It took me about 10 minutes to get the technique down, so you'll have little trouble.

Decals. The decals on this model are very thick, which makes them almost impossible to apply to the hull with its thick panels and deep depressions. The insignias are very simple to paint, so that's what I did. If you choose this course of action, follow the decals as a guide.

Final steps. Once the model is completely dry, there are a couple of final steps. First, you must add the windows. The instructions call for the windows to be black. Since the model is so dark anyway, I decided to paint them white as if backlit. To do this, use camouflage gray on a fine-tip brush to paint them and then add a wash to tone them down. Then drybrush over them to clean up the edges. Finally, apply a coat of thinned flat coat over the whole model and let dry overnight.

The base. To complete the model, I added a star background to the base by cutting a piece of Plexiglas and painting it gloss black. Once it was dry, I dipped a toothbrush in camouflage gray and sprayed the stars by dragging my thumb along the toothbrush.

Shawn Marshall's brilliant *Enterprise* is packed with wires and lights.

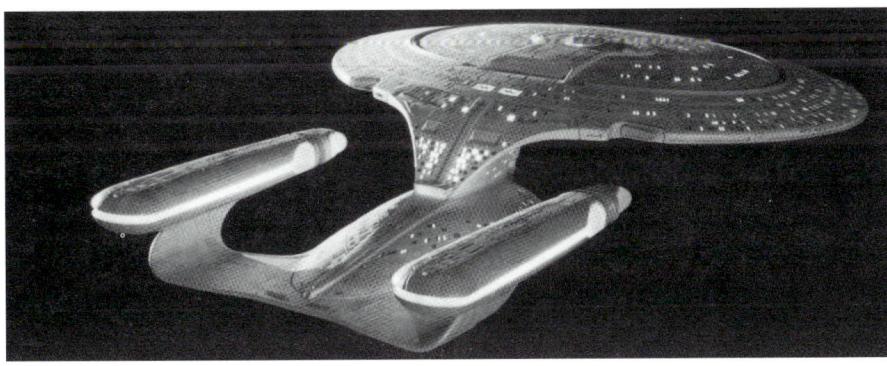 

Studying videotapes of the television show *Star Trek: The Next Generation* gave Shawn lots of bright ideas for superdetailing the Ertl kit.

WIRING ERTL'S "NEXT GENERATION" USS ENTERPRISE

Illuminating a starship with fluorescent tubes and LEDs

SHAWN MARSHALL

Fans of television's *Star Trek: The Next Generation* know the story of the *USS Enterprise*. The 643-meter-long *Galaxy* class starship was built at the Utopia Planitia Fleet Yards above Mars and commissioned on October 4, 2363.

In this particular case the "real" ship is the model, the rest is make-believe. Several models have been used on the TV show. The original was a 6-footer seen during the opening credits and in shots used during episodes. At the end of the third season a 4-foot model was built that featured more surface detail. Modifications included the phaser banks, Bussard ram scoops, and running lights.

51

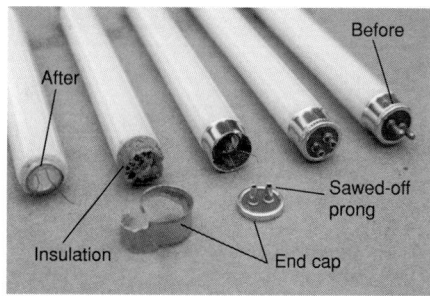

Fig. 1. Here is how Shawn disassembled the fluorescent tubes, the first step in fitting them aboard the Enterprise. Warning: taking these tubes apart may produce fine shards of glass; wear eye and respiratory protection.

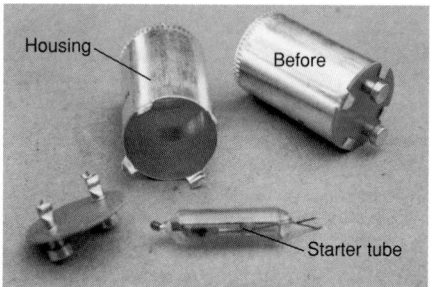

Fig. 2. Shawn also saved space by taking the starter tube out of its housing.

Fig. 3. The tubes' ballasts are far too big to take aboard. Instead, they're concealed in the display base, as is the transformer that runs the LEDs. The eight-pin DIN plug will be fastened at the top of the display stand.

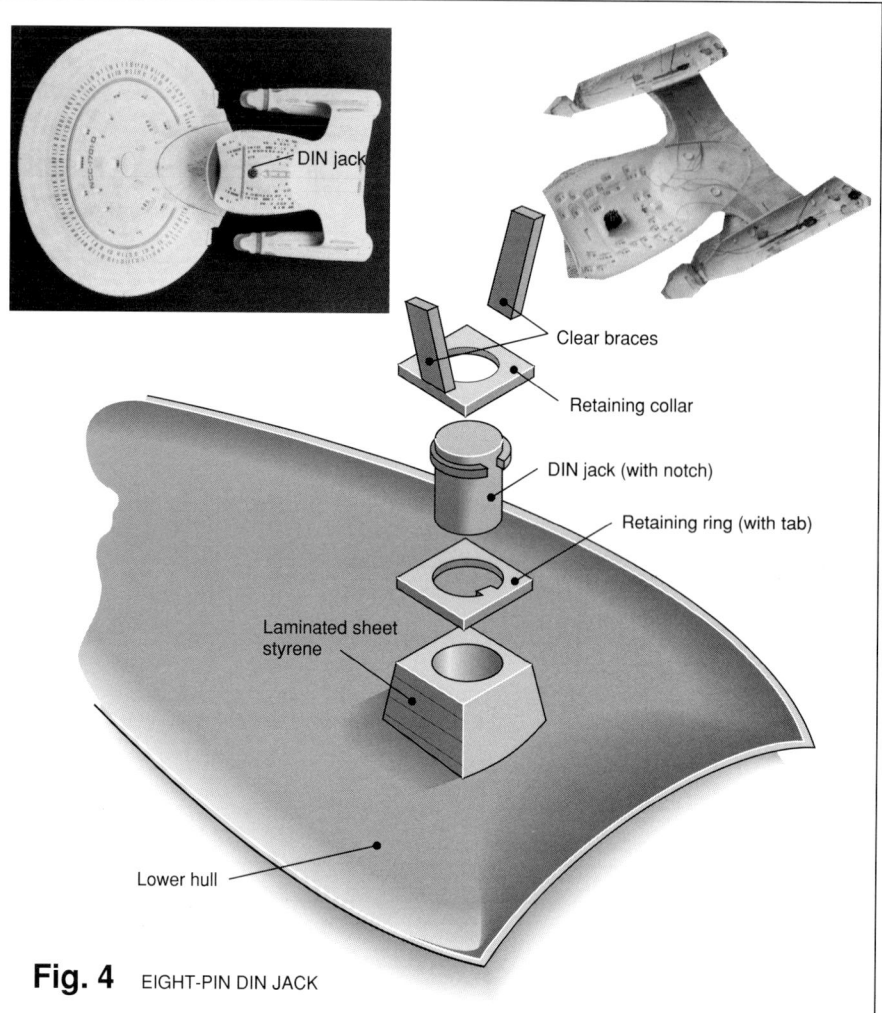

Fig. 4 EIGHT-PIN DIN JACK

My goal was to make Ertl's 18"-long "Next Generation" *Enterprise* (kit No. 6619DO) look like the 4-foot model in the show, despite the fact that the kit is not even half that size.

The plan. My goal was to light more than 1,000 windows as well as the warp engine nacelles, main deflector, main impulse engine, Bussard ram scoops, and running lights. I used LEDs (light-emitting diodes) for the running lights; fluorescent tubes provide the interior lighting.

Fluorescent tubes run brighter and cooler than incandescent bulbs, and seemed especially suited to replicate the warp-nacelle lights. There are five in the ship: one in each engine, another running diagonally through the hull, and two in the saucer section. I used Philips 6" F4T5/CW tubes, rated to last 6,000 hours; I didn't want to have to rip everything apart to change them.

Batteries are too wimpy for a job this big. Alternating current (AC) powers the tubes, which are wired in parallel. A 12.6-volt step-down transformer converts the AC to direct current (DC) to run the LEDs, wired in series. On page 00 you'll find a wiring schematic and drawing to show where everything goes, as well as a parts list. Keep the drawings handy as we go along.

Before we proceed, I must warn you not to try to wire anything without proper tools and training in 110-volt AC connections. Also, taking apart fluorescent tubes can produce fine shards of glass: Always wear protection for your eyes and respiratory system.

Off with the caps. Unmodified fluorescent tube fixtures would be too big to fit in the ship, so I had to separate their main components and wire them back together. I removed the end caps from each tube, Fig. 1. Sawing off the tips of the prongs on each cap freed the delicate wires running from the end of the glass tube to each metal prong. Then I sawed off the end of the cap, peeled back the metal wrapped around the end of the tube, and gently scraped the brown insulation off, careful to avoid damaging the glass seal.

Each tube requires a starter, a glass tube in an aluminum housing, Fig. 2. The housing is too big to fit inside the ship, so I bent back its sides, clipped the wires from the

WARNING! Shocks from improper handling of 110-volt alternating current can cause serious, possibly fatal, injury! Do not attempt this project unless you have proper tools and training in making 110-VAC connections, or are assisted by someone who does.

USS *ENTERPRISE*, dorsal view **Not to scale**

PARTS LIST

- Yellow LEDs: Cal Pak (part No. CP-35), Cal West Supply, Westlake, CA 91362

The following parts are available from Radio Shack stores:

- AC-to-DC step-down transformer (part No. 273-1385)
- Brilliant red LEDs (part No. 276-087)
- Rectifier (No. 276-1146)
- 555 timer chip (No. 276-1723)
- 7805 voltage regulator (No. 276-1770)

USS *ENTERPRISE*, wiring schematic

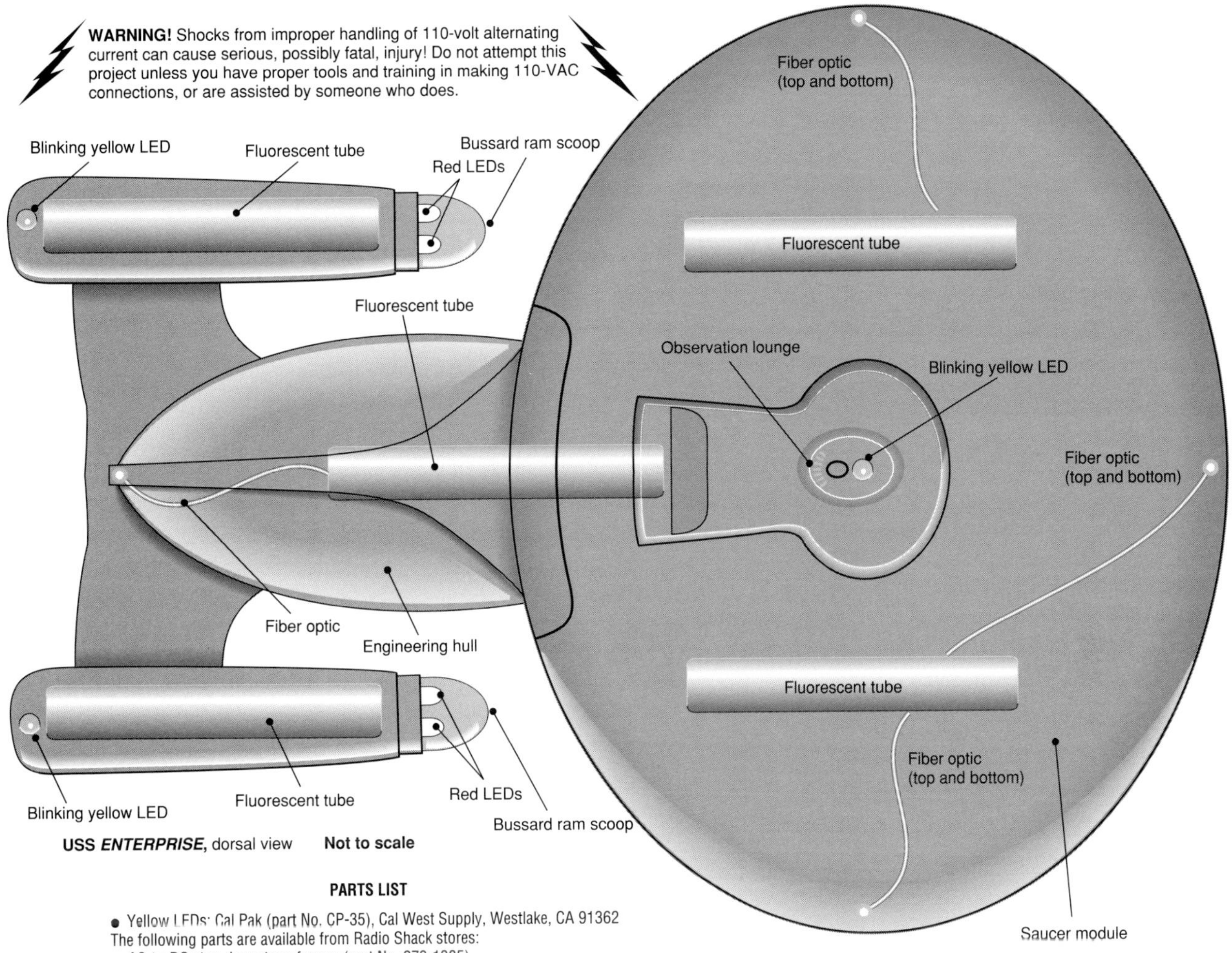

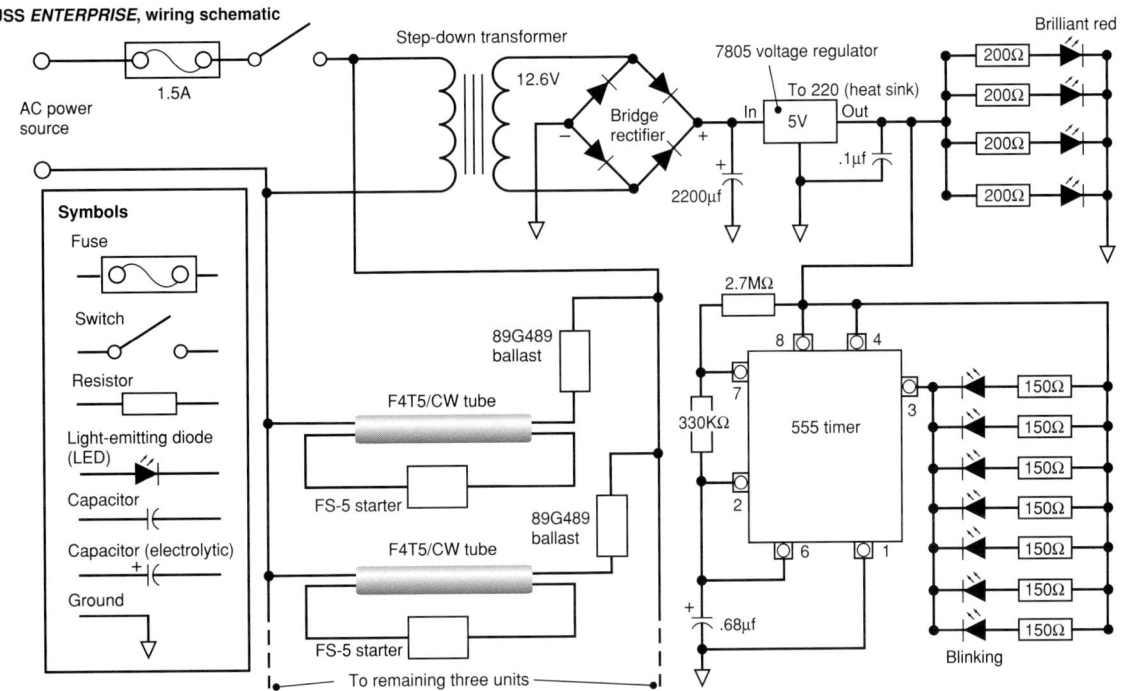

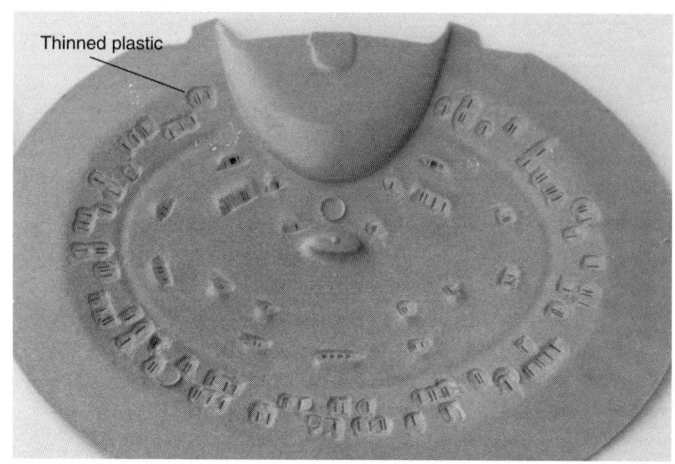 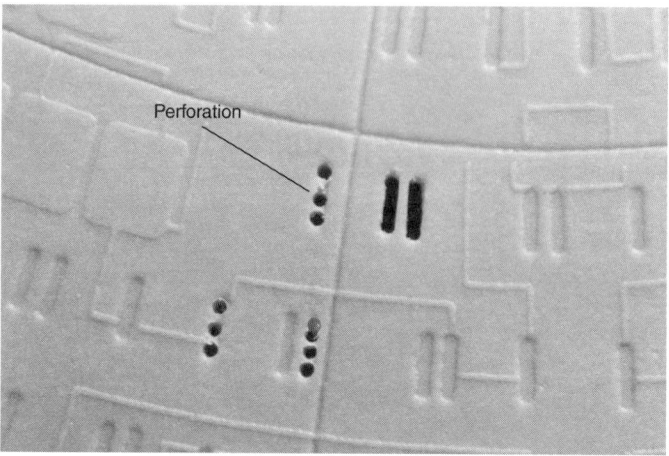

Fig. 5. Shawn marked the windows, thinned the plastic from the inside, perforated the window openings, then cut them out.

base, and wired each starter directly to its tube.

Another component of each light is the ballast, which regulates the voltage that flows through the tube. The ballasts are far too bulky for the ship: I concealed them in the model's display stand, Fig. 3.

Connections in the stand. An eight-pin DIN junction provided the minimum number of connections needed. I ran wires for the plug (male) to the top of the stand, Fig. 3, and mounted the jack (female) in the bottom of the lower hull, Fig. 4.

The jack is housed in a sheet-styrene lamination shaped to the hull's contours. After gluing the housing just aft of the bottom phaser bank, I flipped the part over and drilled a hole for the jack, making sure the hole was perpendicular to the hull. I glued the jack in the hole with super glue, attached a tab-in-notch retaining assembly, and added clear-styrene braces running diagonally from the top of the jack housing to the underside of the upper engineering hull where it meets the dorsal hull.

Lights in the windows. Figure 5 shows how I made each of the 1,013 windows. I drilled a hole from the outside to mark a window location, then thinned the plastic from inside with a motor tool. After thinning the plastic, I drilled the window outline to create a perforation that was easy to cut and smooth with a hobby knife. I painted the interior to prevent light from bleeding through the hull, spraying it black, then silver, then gloss white.

Later, after painting the hull exterior I installed .010"-thick clear styrene windows. A light sanding frosted the clear plastic, diffusing the light for a more realistic effect.

Nacelles and their lights. Each nacelle contains one fluorescent tube, two yellow blinking LEDs for running lights, and two brilliant red LEDs to replicate the glow of the ram scoops. I chose LEDs over grain-of-wheat bulbs because of their longer life.

Powering these lights meant running five wires to each nacelle—difficult because the nacelle supports are solid plastic. I chose a nearby panel line to scribe a groove deep enough to accommodate the wires, labeled each wire, super-glued them in the groove, and covered them with filler putty. After sanding the area smooth I restored the original panel line with thin wire, Fig. 4.

I drilled a hole for each LED and countersank them from inside with the motor tool. After gluing them in place I painted their backs black to keep the light localized, Fig. 6.

Fig. 6. After drilling a hole to fit the LED, Shawn back-painted the area black to conceal the shape of the light.

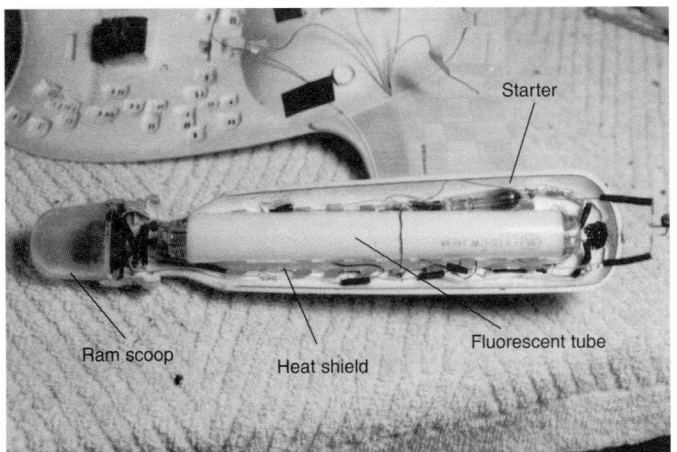

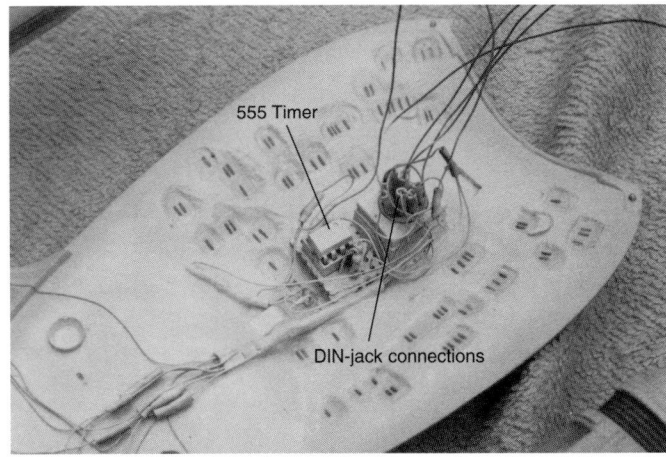

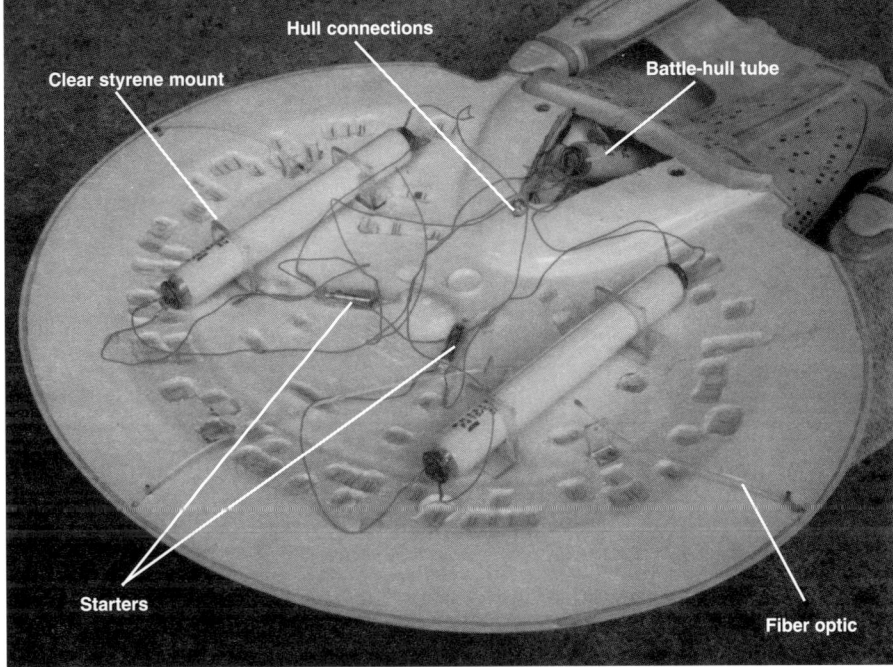

Fig. 7. Close quarters: You can see why the tube had to be broken into its components. An aluminum heat shield protects the plastic.

Fig. 8. The timer chip synchronizes the blinking lights. Note the shrink-tube insulation on the DIN jack.

Fig. 9. You can see one tube emerging from its diagonal course through engineering and the battle hull. There's lots more room to work with in the saucer module.

For the two red LEDs inside each ram scoop, I made mounts from .025" sheet styrene to hold them upright inside the scoops. Spraying scoop interiors clear flat and painting the backs of the lights black concealed their shape.

Although the fluorescent lights run cooler than incandescents, the ends of the tubes are still too hot to touch. I made heat shields from .010" aluminum sheet and fixed them inside the nacelles with epoxy. I wired each tube and secured it to the lower heat shield with wire, then epoxied the starter to the heat shield, next to the tube, Fig. 7.

A theatrical tint. Theatrical stage lights are colored with a thin, heat-resistant tinted sheet of acetate called a gel, available from photo supply stores. I super-glued a strip of Roscolux No. 67 (light sky blue) gel inside the nacelles. The heat-resistant gel gives the warp engines their distinctive blue glow. I glued and sealed the nacelles, then softened their molded detail with 320-grit sandpaper. I also added grooves around the ram scoops at the front, drawing them with a pencil and cutting them with a razor saw.

Blinking in synch, down in engineering. A 555 timer chip blinks the running lights in unison, Fig. 8. I mounted this small circuit just aft of the stand connection, painted it gloss white, then soldered each connection and covered it with heat-shrink insulation, checking every light at each step to ensure that nothing had shorted or disconnected.

I seated a 15-ounce lead weight in the hull to counter the weight building up at the power connection, shaping the weight to the hull so it wouldn't block light from the rear windows. Painting the lead gloss white made it reflective.

Battle hull and saucer. After gluing the dorsal-hull halves in place, I fitted the battle-hull fluorescent tube tightly in a scrap of clear sheet styrene and ran it diagonally through the dorsal and engineering hulls, gluing its starter forward of the stand connector. Cutting through the top of the battle hull and into the rear bottom of the saucer eliminated a main connection point on the model, so I glued and bolted the bottom of the saucer to the battle hull, then cut the hole for the tube with a motor tool.

An .060" fiber-optic strand runs from the tube to the rear of the engineering hull for a light above the aft photon torpedo tube. I heated the end of the fiber-optic strand to create a mushroom-shaped head, drilled a hole in the hull, and fed the strand through from the exterior. After gluing the fiber-optic head to

Fig. 10. The observation lounge. Look closely—there's furniture in there!

Fig. 11. (Top, right) Here's a switch: The airbrush provides the atmosphere, the turntable does the work. Shawn spun the planet while painting it.

Fig. 12. The completed stand. Shawn stuck a brass tube in the planet, fastened the plug in the top, and added support to keep *Enterprise* on an even keel.

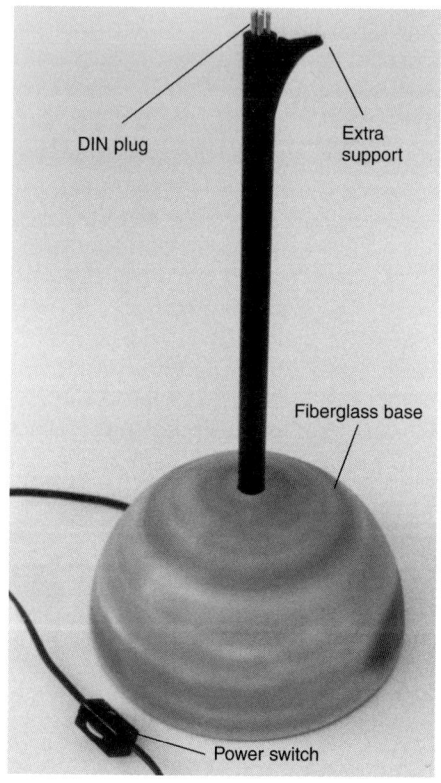

the hull, I glued the lower engineering hull and warp-engine assemblies to the upper engineering and dorsal hulls, Fig. 9.

I mounted two light tubes in the saucer with clear styrene scraps (from an old toothbrush box) and attached the starters inside the hull with epoxy. More .060" fiber-optic lines replicate non-blinking lights port, starboard, and forward. I tinted the port and starboard fiber optic lines red and green, respectively, with stained-glass paint (a lacquer-based clear tint available from stained glass studios).

Saucer. To add a bay of six windows just forward of the captain's yacht on the underside of the saucer, I cut out a square hole, beveled the edges, and installed a bay made of scrap sheet styrene. Gluing Roscolux No. 26 light-red gel behind a clear part on the impulse engine gave it a red glow.

I cut out seven windows for the observation lounge aft of the bridge, Fig. 10. I didn't frost this because on the show you can see people walking around inside as the ship roars past the opening credits. A tiny table, floor, and back wall in the lounge add dimension to this detail.

I installed the blinking LEDs just forward of the bridge, fixed a blue gel behind a set of square windows aft of the shuttle bay, then glued the saucer halves together.

Painting. With so many subassemblies I painted the ship as I went along, starting with the engines and moving to the engineering and dorsal hulls before arriving at the saucer. I compiled a videotape reel of *Enterprise* footage for reference.

To paint the complex mosaic of hull panels I made a stencil by laying parallel, slightly overlapping strips of tape on the hull and rubbing a pencil over the raised detail. I transferred these strips to wider strips of clear packing tape lightly stuck to a piece of glass, and cut out the patterns with a hobby knife. The same technique produced a stencil for the nacelles.

I airbrushed a base coat, a mix of gloss white, black, and blue enamel, then applied the stencil to the hull and airbrushed a slightly darker tone of the base coat. After stenciling the starboard half of the saucer section, I was able to use the leftover stencils (still on the glass) by giving them a light coat of spray adhesive, flipping them over, and tacking them on the port side of the saucer section.

More than 400 lifeboats ring the saucer section and lower engineering hull. I made stencils for them and painted them light tan. The phaser banks are medium dark gray, the gas filters behind the ram scoops orange.

Where possible I masked and painted details, preferring paint to decals on the intricate hull. I did use the decals for insignia and markings, applied them with a generous dose of white glue and decal-setting solution.

Oddly, there are no molded windows on the outer edge of the upper saucer half. I made a pencil rubbing of the lower half windows, cut a stencil, and airbrushed the missing windows black. In the same way, I added windows to the edge of the engineering hull.

I airbrushed a thin black for light weathering and cleaned it up a little with slightly darkened white. The remaining, unopened windows are filled in with a black Rapidograph drafting pen. I coated the completed model with clear flat.

Finishing the stand. Posing the *Enterprise* in orbit allowed me to create a planet to house the wiring. I made a fiberglass hemisphere, shaping it on a mold made from an old lamp base. A hole in the top of the planet supports a brass tube; L brackets hold it to the base plate.

To paint an atmosphere on the planet, I spun it on an old turntable as I airbrushed cloud patterns with shades of blue and white, Fig. 11, finishing with clear flat.

Mounting the ballast. The five fluorescent-tube ballasts and the AC-to-DC transformer are mounted on a 3/32" sheet-aluminum base plate; rubber feet prevent slippage. I ran the wiring through the brass tube stand, wired the DIN plug, and fixed the plug in the top of the stand with JB Weld, a metal adhesive and filler available at hardware and auto parts stores, Fig. 12.

I was concerned that the ship was nose-heavy, so I fastened an extra support to the stand. A piece of black felt pads the curved part of the support that touches the hull; the support is attached with JB Weld.

I painted the tube and support black semigloss, screwed it to the display base, and installed a switch in the AC cord.

Engage! With great anticipation I plugged in my *Enterprise* and hit the switch. The fluorescent lights flickered on with that distinctive crackle, the Bussard ram scoops glowed bright red, the running lights blinked in unison, and 1,013 windows twinkled. My *Enterprise* was complete—and it looked great!

Chris's 1/500 scale *Enterprise* boldly goes where no model has gone before! Forsaking the kit's engraved panel detail, the author painstakingly airbrushed the ship's "Aztec" pattern with frisket film masks. Color photos by Chris Becker.

ACCURIZING ERTL/AMT'S STARSHIP ENTERPRISE
Constructing a movie-worthy Star Trek miniature

CHRIS PAVEGLIO

One of the most relaxing segments of this hobby is modeling spacecraft. Compared to aircraft, autos, or armor, science fiction miniatures require little, if any, research. Space vehicles can be built in any scale, made from anything—including as much imagination as you want.

But the universe of *Star Trek* is different: There is a purpose for everything, so you can't just call a golf ball a fuel tank. Ertl/AMT's movie *Enterprise* kit is a good model, but good doesn't necessarily mean accurate. I hope this chapter serves as a shortcut for building your own accurate movie *Enterprise*.

I've divided the project by ship sections, but you may want to do all the major puttying and sanding at the same time. Remember to wash all parts with soap and water first.

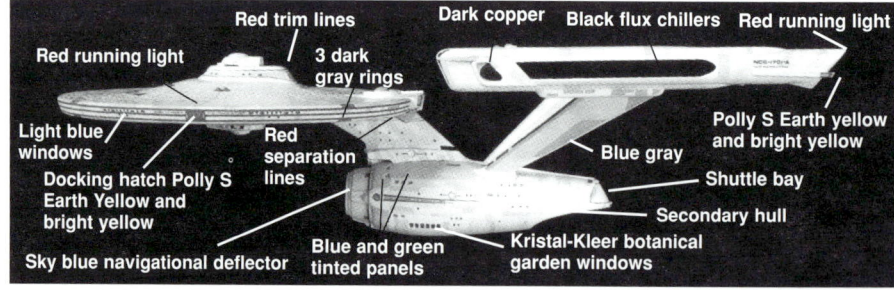

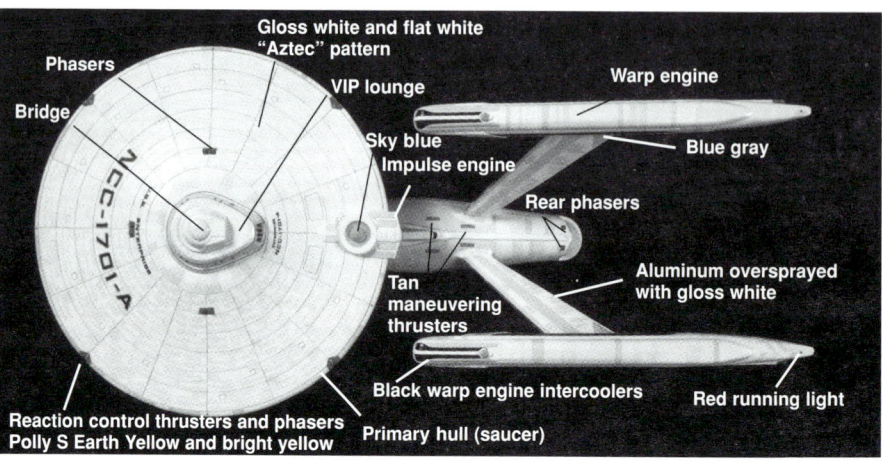

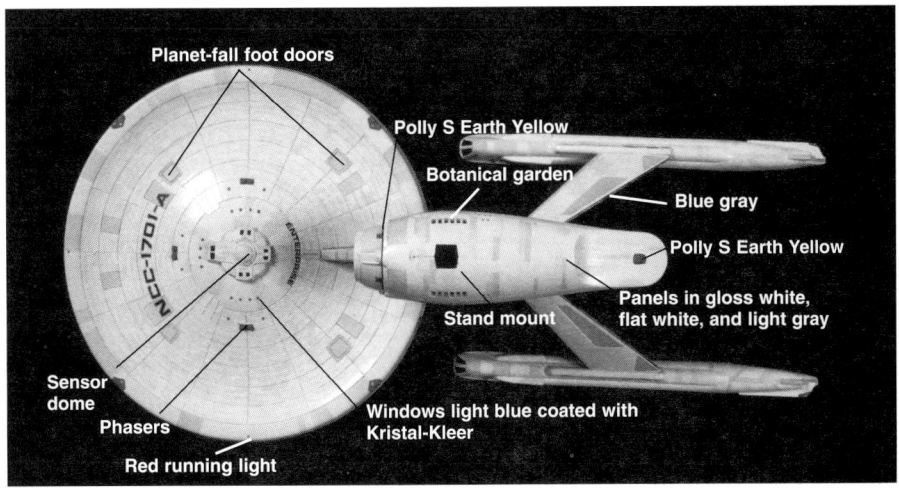

The best references on the movie *Enterprise* are the movies themselves. *Star Trek* and *Star Trek II: The Wrath of Kahn* are excellent visual guides for the tonal variations in the paint scheme. They are also fun to watch!

The primary hull (saucer). Remove all flash, alignment pins, and the large bump behind the bridge on kit part No. 3. Remove the bridge dome with a razor saw, using the middle ring as the slicing point. Sand the top of the bridge, then file the bottom of the dome until one ring and the four squares remain. Reattach the dome with the squares facing as they were before removal (Fig. 1).

Cut the sides and top of part No. 4 so that it will fit inside part No. 3, then glue it in. Fill the depression on both sides of the circular docking port and sand smooth. Sand off the raised panel lines on the sides of the bridge. Using a 5/32" drill bit in a motor tool, slightly enlarge the docking port to make it the same size as the others on the ship.

To flatten the rear face of the VIP lounge on the top of the primary hull (part No. 1), file it down until it no longer appears swollen. Using a small file, enlarge the four vertical windows, leaving thin frames between them. Enlarge the eight circular and two oval windows around the lounge with a small file, then back them with bits of styrene or putty. To be fancy, insert a sheet-styrene floor for the lounge, and try to make tiny furniture from bits of styrene!

Cut off the phaser and running light bumps on the top and bottom halves of the saucer—these will be replaced later. Relocate the lower left window on each of the six parts No. 6 (Fig. 2), then clean and slightly enlarge each window with a needle file. Open and clean the engraved windows at the left rear of the edge of the saucer bottom. The corresponding windows on the right rear edge are incorrect, replace them with eight rectangular windows to represent the recreation room (Fig. 3).

Now comes the fun part. The recessed random panel line pattern on the kit's parts is incorrect. First, you'll need to fill in the kit pattern with putty. (There is a definite pattern on the saucer we'll reproduce when we paint the model.) The heavily engraved radial and concentric grid lines remain, along with the hatches, docking bay doors, and planet-fall foot doors, but there's a catch. Ertl/AMT didn't mold all the hatches and doors into the bottom of the saucer. Engrave new access panels, foot doors, and docking bays as shown (Fig. 4).

Before assembling the saucer halves, fill the lightly engraved panel detail on both surfaces with filler putty, then immediately clean out the heavy radial and concentric lines and all existing hatches with a needle. Cover a small section at a time, then clean out the lines. Don't fill the detail at the circumference of the saucer halves right now; just take care of the "dished" areas.

After the putty has dried overnight, sand the surfaces with 320-grit wet-or-dry sandpaper and water, followed by 400- and 600-grit. After sanding, clean out the heavy engraved lines with the needle and lightly sand again. Your goal is a smooth surface with only the heavy engraved lines showing.(Fig. 5).

Use a drill bit in a motor tool and a needle file to clean out the 18

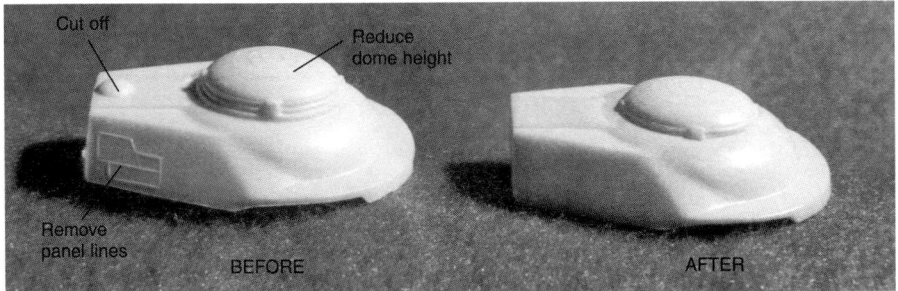

Fig. 1

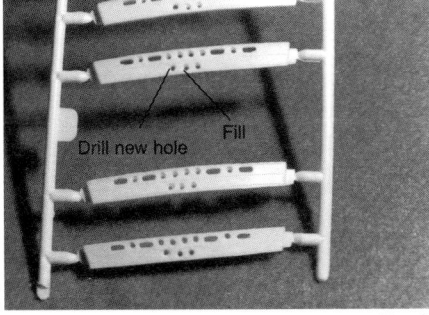

Fig. 2

Fig. 3

windows on the bottom of the saucer; you can find them from the inside. Back the windows with bits of styrene sheet.

Cement the six window bays (part No. 6) into the edge of the saucer bottom, then glue the saucer halves together. Once the cement is set, putty the seams and the panel detail on the circumference of the saucer; take care not to putty over the windows or the reaction control thrusters. When the putty is dry, file and sandpaper the surface smooth. Keep the circumference flat—don't round off the corners too much.

With a knife, lightly scribe lines across the flat faces of the impulse engine exhaust (part No. 7). You may need putty and sandpaper to get this part to fit into the saucer.

To refine the sensor dome (part No. 5) attached to the bottom of the saucer (Fig. 6), use a knife and files to reduce the size of the housing protrusions, then a routing bit in a motor tool to enlarge the vents. Thin the edges of the four scoops with a knife and a file. Don't attach the sensor dome to the saucer just yet, but you can attach the bridge to the VIP lounge.

Secondary hull and warp engines. Let's start with what I call the neck; it connects the primary and secondary hulls. Extend the forward edge of the neck with sheet styrene and putty (Fig. 7). When mated to the saucer, this extension will reach the innermost concentric grid line.

Putty over the recessed rectangular panel detail as you did on the saucer, but keep putty out of the

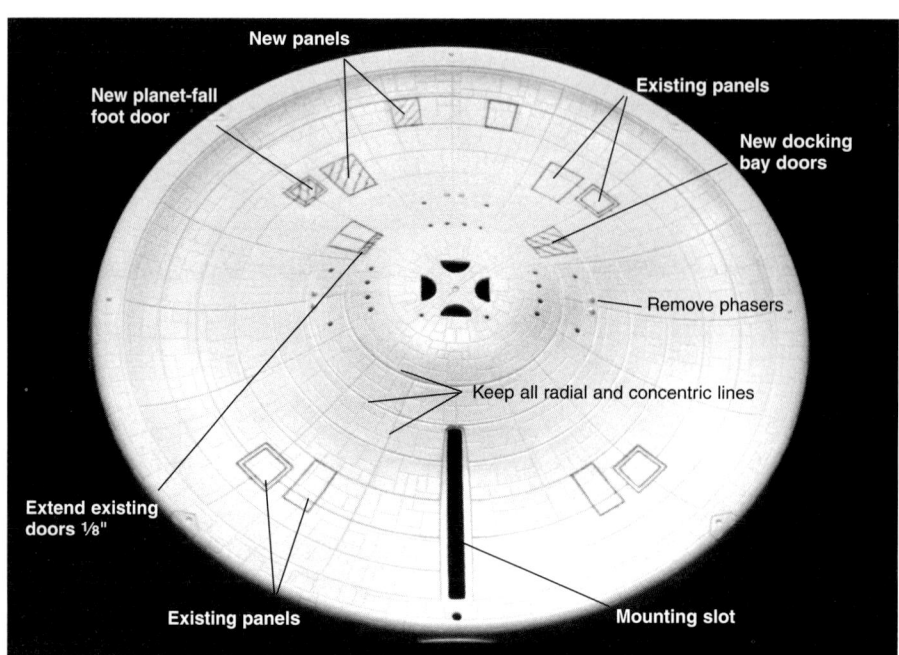

Fig. 4

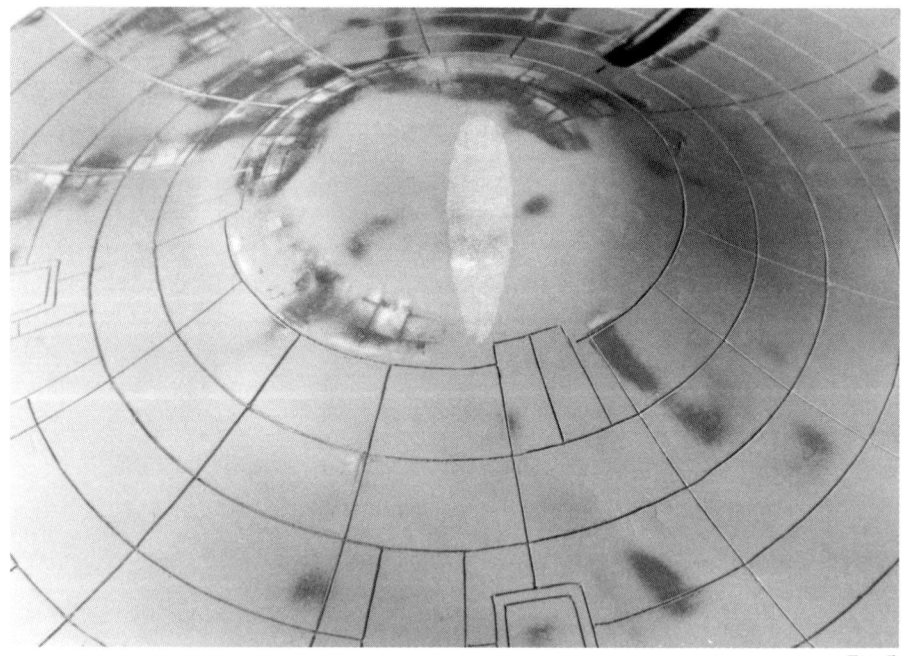

Fig. 5

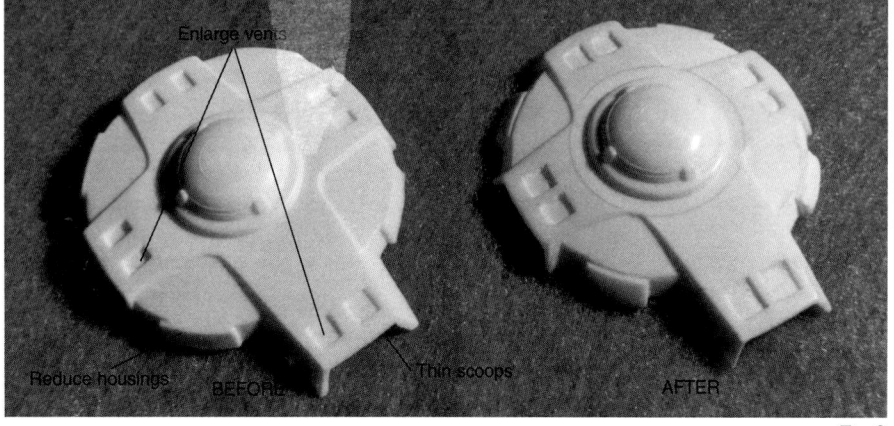

Fig. 6

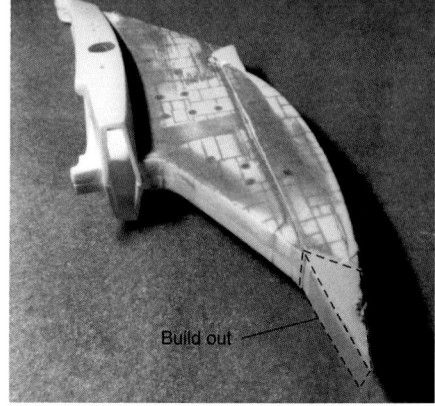

Fig. 7

Fig. 8

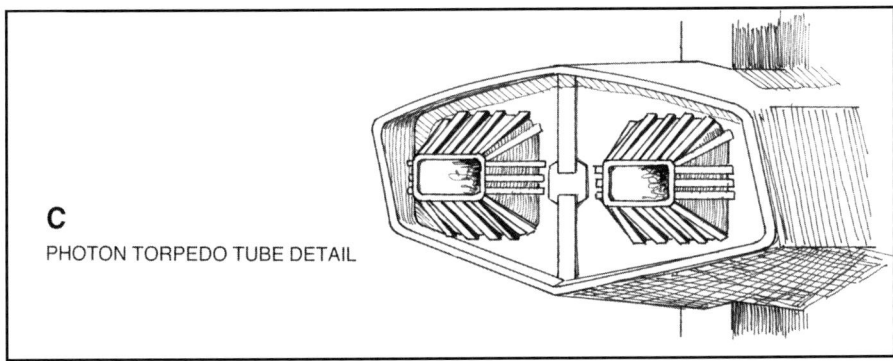

Fig. 9

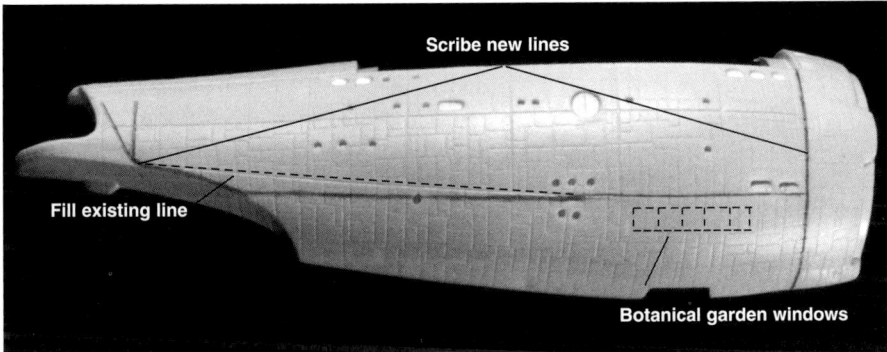

Fig. 10

photon torpedo exhaust at the bottom rear. Clean out the separation line and the windows, too.

Using the patterns provided, make two new impulse engine intercoolers from .010" sheet styrene (Fig. 8) and scribe horizontal lines with a scriber or blade, then glue the new intercoolers to the top of the neck.

Build up the photon torpedo tubes (part No. 10) with putty and stretched sprue. The finished tubes should look busier than the kit part (Fig. 9).

Assemble the secondary hull as in step 1 of the kit instructions. For the botanical garden, use a drill bit to open up the two six window sets near the bottom of the hull, then carve them so you create thin frames between the panes.

Putty over the panel detail once again. Keep clear of the engine pylon mounts and the eight maneuvering thruster depressions on the top: clean out the windows and docking ports. Scribe a new rear bottom grid line on both sides of the hull (Fig. 10) and a grid line around the front of the hull. Glue the shuttle bay doors in.

Once again, fill in the engraved rectangular surface detail on the engine pylons, but keep the ribbed vents clean. The same goes for the warp engines, leaving the front intercoolers and the flux chillers along both sides clear of putty. Carve off the three tiny bumps at the tail of each engine and take your time fitting the front intercoolers to the engines.

Make sure the neck is perpendicular to the secondary hull. Also, be careful of the alignment of the warp engines on their pylons.

Painting panel detail. The starship *Enterprise* is not painted just white; it's painted many shades of white. After studying the movies, I found that the random pattern engraved in the kit is completely unlike the panel pattern on the "real" *Enterprise*— itself a model. That's why I eliminated the random pattern and prepared the model for a more accurate paint job.

Keep the saucer separate from the rest of the ship; it'll be easier to paint that way. First paint all the subassemblies with a light gray primer and inspect them for blemishes, seams, and other surface irregularities.

Most of the ship is airbrushed with three shades of white: gloss white with some silver added; flat white; and flat white darkened slightly with medium gray. Paint various panels on the secondary hull and neck with green and blue and overspray them with a light coat of gloss white, or tint the gloss white with green and blue. I oversprayed my *Enterprise*. Paint other panels with light blue-gray or aluminum.

Remove the tape from masked panels as soon as the paint is dry to the touch. Tape left on for more than a day or so could leave adhesive behind, making an awful mess. Study the color three-view photos on pages 54 and 55 for painting information.

Painting the primary hull (saucer) is time-consuming, but the finished model looks great! Make masks for the saucer's "Aztec" pattern from frisket film (large sheets of self-adhesive film available in art supply stores). You can cut frisket film with a sharp knife and reposition it frequently. The Aztec pattern is complicated; apply each section individually. You can see that the pattern of flat white and gloss white alternates around the saucer.

First paint the saucer gloss white (with some silver added) and let it dry thoroughly. Cut out the areas shown in white on the templates (Figs. 11 and 12), and apply them to the surface. Use the remaining portions of the frisket (shown white on the template) to mask the next grid section over. Duplicate

Fig.11 & 12

Fig.13

this process repeatedly as you work around the saucer. Now spray the saucer flat white. When the paint is dry, remove the frisket masks to reveal the gloss white.

Paint the navigational deflector in the front of the secondary hull bright sky blue, then use a curved blade to scrape the blue paint off the radial lines. Use sky blue on the impulse deflection crystal at the rear top of the saucer and the small domes at the top front of the warp engines. Paint all the small windows light blue, then apply a drop of Micro Kristal-Kleer to simulate glass.

Cut off all but 1/32" from round-headed map pins and replace the kit phasers. Grind depressions in the phaser locations with a routing bit in a motor tool, then glue the pin heads into the depressions. Paint phasers and reaction control thrusters with Polly S earth yellow and bright yellow details. Paint three dark gray rings around the circumference of the saucer, then apply tiny dots of red for the running lights on the saucer and warp engines. Now you can epoxy the saucer onto the neck; make sure it sits perpendicular.

Place the saucer's large NCC-1701-A markings in the third grid ring of both the top and bottom surfaces of the saucer. (The instructions show them on the second ring). The ship name beneath the shuttle bay doors is missing from the decal sheet. I used decal No. 7 from a spare sheet, cutting between the letters so the decal would fit on the curved surface (Fig. 13). One each of decals No. 5 and 6 should straddle the docking ports on the port and the starboard sides of the saucer. Use thin red decal strip to outline the VIP lounge ribs and the separation line on the neck just below the saucer.

REFERENCE
• Newitt, Paul Matthew, *Star Fleet Assembly Manual Four*, PMN Designs, Davis, California, 1983

The author reworked the interior and stern of Ertl's *Galileo* shuttle craft. He estimates it's about 1/35 scale.

MAKING GALILEO ACCURATE
Recalibrating the Galileo of the original TV series

MARC MILLIS

The last anyone saw of the of the *Star Trek Galileo* shuttle craft was in the original TV series—Stardate 2821.5, to be more precise. That was when Spock signaled USS *Enterprise* from the shuttle and was beamed aboard moments before *Galileo* burned up in the atmosphere of Taurus II.

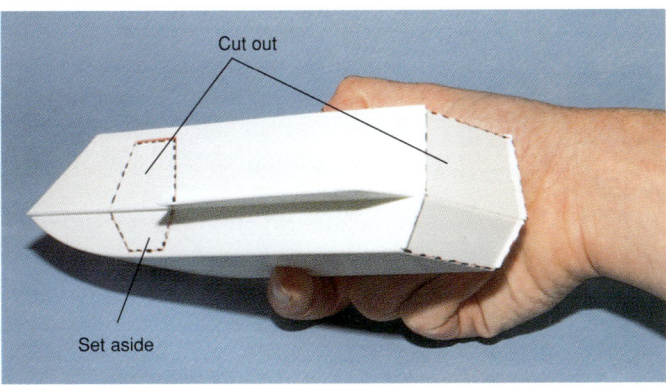

Fig. 1. Renovation starts with demolition: Hatch and back-wall cutouts are marked with dotted lines. Dick McNally photo.

However, the Ertl model *Galileo,* and reissue *Galileo II* (kit No. 6006), have reappeared since then. Although it's not in production now, it's still available in stores—and chances are it will visit our galaxy again.

Sizing up inner space. *Galileo* was bigger inside than out. No, this wasn't some futuristic warp technology—only Hollywood expediency. Shooting scenes inside the craft required more elbow room than a shuttle would provide.

The Ertl kit isn't labeled with a scale, but I decided the model is 1/35 scale. That's about right for the seats the kit provides, and it agrees with the plans in Franz Joseph's book *Star Fleet Technical Manual*.

Open the door, put up new walls. I began my modifications by cutting out the main hatch and the aft walls, Fig. 1. Save the hatch's lower section.

To replicate the interior sliding doors seen in the TV episodes, I erected interior walls of .030" sheet styrene, Fig. 2, and inserted .030" sheet between them and the outer hull to replicate retracted doors, Fig. 3. I glued the lower-door section in a folded-down position.

Of course, the walls had to include window cutouts. I laid the sheet styrene over the inside of the windows

63

Fig. 2. Looking at the open hatch reveals the double hull.

Fig. 3. Open door: The kit hatch section is folded down, and sheet styrene inside the walls suggests a retracted door.

Fig. 4. Curling sheet styrene curves the ceiling. That light is from a wristwatch package.

and scribed the openings from the other side.

I replaced the kit windows with .010" clear styrene sheet, Fig. 4. I glued one sheet to the inside of the kit hull and another inside the new inner wall, then glued the inner wall and window in place. This double pane corrects the window recesses.

A scrap of clear plastic from a watchband package replicates the ceiling light. The plastic had the desired grid pattern; I simply cut it to length and glued it in place. Spraying it with Dullcote frosted the "glass." I curled ¼"-wide strips of .010" sheet styrene for the curved ceiling.

To suggest a closed sliding door on the back wall, I scribed a vertical line in the .030"-sheet wall, then cut a portal in .015" sheet and glued it to the back wall, Fig. 5. Scraps of .010" sheet depict control panels on this wall.

Recalibrating the instruments. I copied diagrams in the *Technical Manual* and built a new control console from .020" sheet styrene, Fig. 6.

Photo film was used on the left and right panels. Scrape the emulsion (flat finish) side of a blackened negative with a razor or hobby knife to expose clear areas in the shape of the instruments, and paint that side of the film. You see only sharply outlined areas of color from the other side. I glued the painted film to the new console.

On the central panel I used a decal from Ertl's *Star Trek* Bridge. I also drilled holes and inserted controls made from 1/16" brass rod.

The "astrogator" (as it's called in the *Manual*) is made from a circular portion of the unused kit windshield, Figs. 5 and 6. Replacing the pencil on a compass with a scribing point, I etched circles and radii in the clear plastic.

Scanners on top and on both sides of the console are represented by 6 mm plastic spheres I found at a craft store. I bored out each sphere, painted a spare car headlight black, and glued it on the sphere. The spheres are glued to hand rails from a Revell Bucket Wheel Excavator

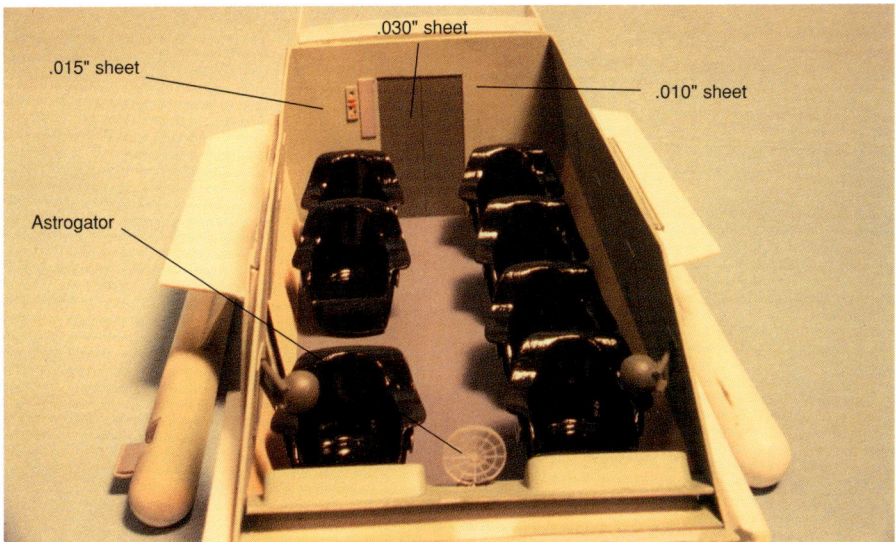

Fig. 5. Front and center on the control console is the astrogator. A .015" sheet frames the scribed door.

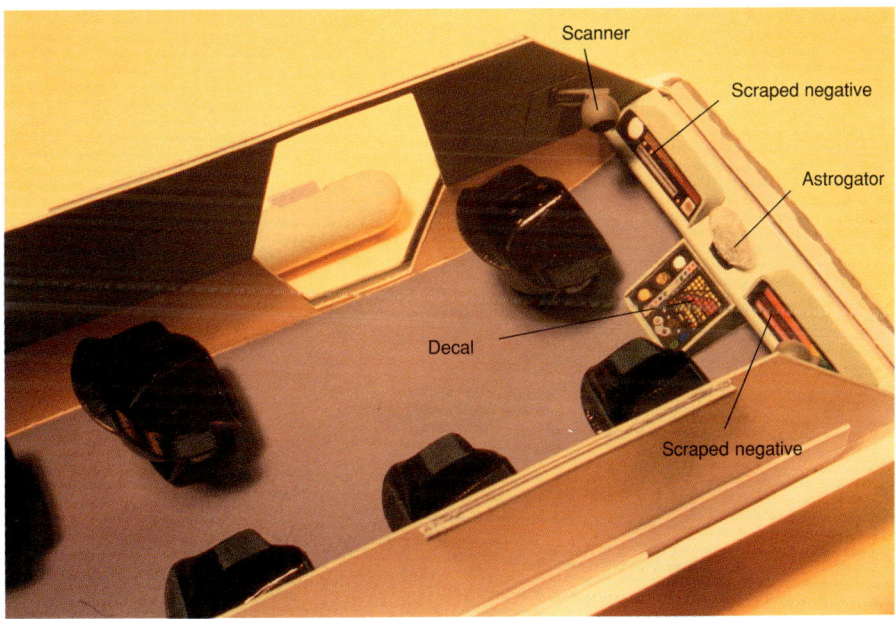

Fig. 6. Controlling the shuttle: Film negatives, scraped and painted, provide colorful readings. The scanners are made from craft beads and car lights.

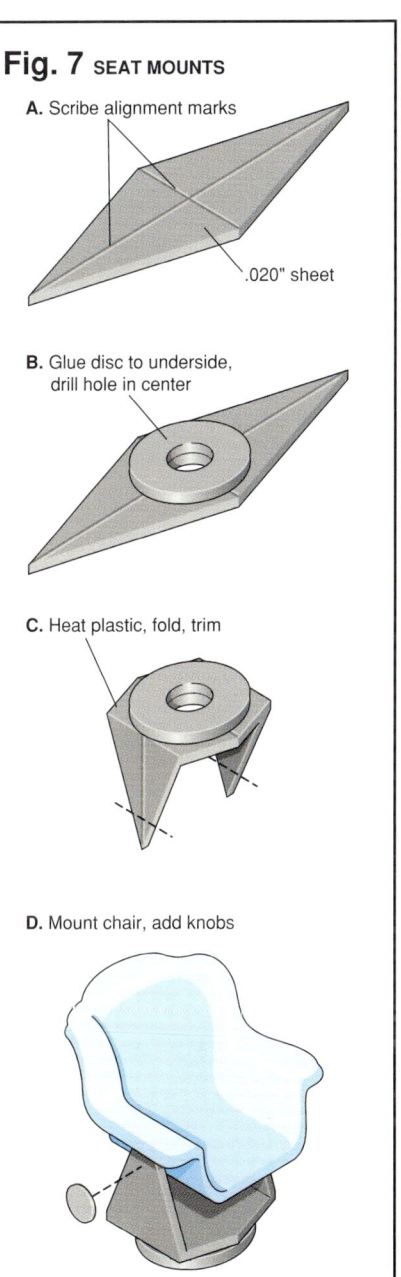

(an excellent source of spare gizmos) and attached to the walls.

Paint the walls, have a seat. I brook no argument over the interior colors, which vary from episode to episode, especially on my TV. I chose pale green, but learned later that pale blue is more accurate.

I was satisfied with the look of the kit's seats except for the mounts. Figure 7 shows how to build new ones.

The mount started as a diamond-shaped sheet of .020" styrene. I glued a disc (also from the Revell Excavator) to the bottom and drilled a hole through its center. Warming the styrene first to make it more flexible, I bent it to shape.

After gluing the mounts to the seats I added plastic discs (again from the Excavator) to the outside of these joints to depict seat-adjustment knobs.

New engine, completed hull, and markings. I rebuilt the stern of the craft to accommodate a more accurate engine, Fig. 8. The main pieces of the angular fairing are cut from .030" sheet. The fairing is trimmed with .015" sheet.

The kit's engine decal served as a guide to building the new engine. I built eight "cells" from .015" sheet styrene and faced the engine with .020" sheet, Fig. 9.

I glued the hull halves together, covered the spaces between the inner and outer walls at the rear of the craft with .015" sheet to present a smooth surface, then glued on the stern. The engine-access panels

65

Fig. 8 GALILEO STERN

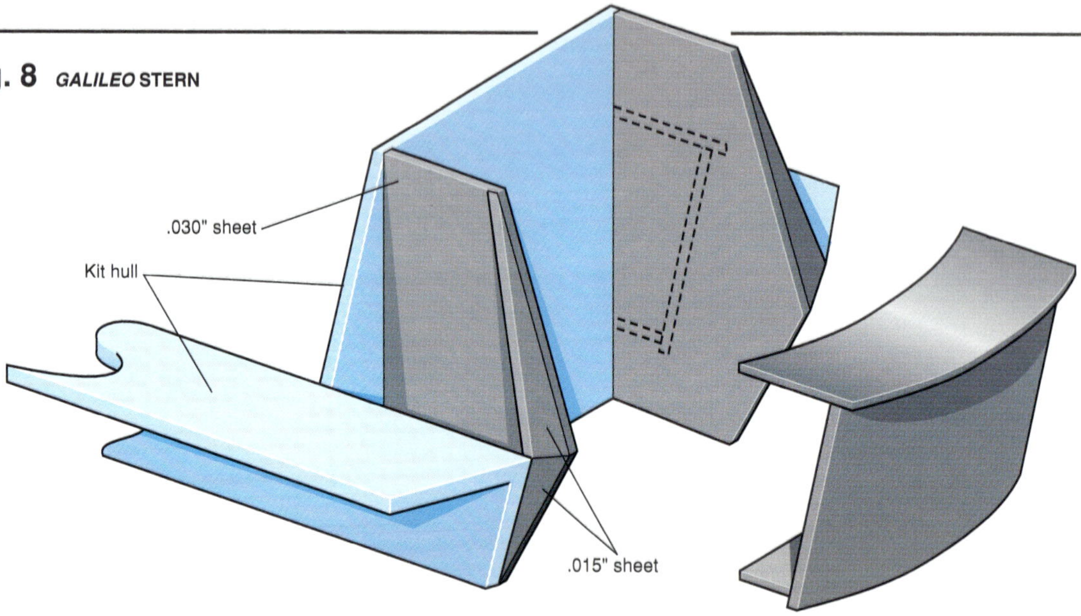

The rebuilt stern holds a souped-up engine

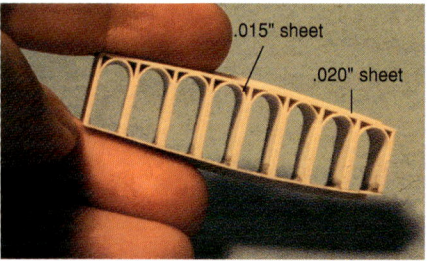

Fig. 9. Engine parts: The curved pieces are .015" sheet styrene. Other parts, including the fascia, are .020" sheet.

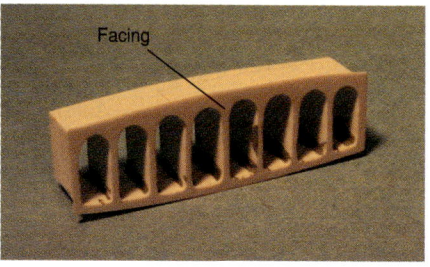

are made from .030" sheet styrene.

A combination of nose and main landing gear from a 1/144 scale NASA shuttle forms the aft landing gear. I dressed up all the landing gear with 1/16" and 3/32" aluminum tubing and added craft-store hemispheres to the kit's landing pads.

I added an identical hemisphere to the rear of each impulse-engine pod, and reworked the step notch on the left engine pod to replicate a fold-out panel, Fig. 3.

To change the decals of the *Galileo II* to the earlier *Galileo,* I simply snipped off the Roman numerals.

Murphy rules the universe. I found excellent references after building the model—that's always the case, it seems. But I'm satisfied that I've made the kit look more like its TV counterpart.

REFERENCES
• *The Making of Star Trek,* Stephen E. Whitfield and Gene Roddenberry, Ballantine, New York, 1975
• *Star Fleet Technical Manual,* Franz Joseph, Ballantine, New York, 1986

SOURCES
• Sheet styrene: Evergreen Scale Models, 12808 N.E. 125th Way, Kirkland, WA 98034
• Brass and aluminum rod and tube: K&S Engineering, 6917 W. 59th St., Chicago, IL 60638

ASSIGNMENT SOUND STAGE

RICHARD G. VAN TREUREN

From the beginning, the platform on which Producer Gene Roddenberry wanted to circulate his *Star Trek* characters on presented a problem. Although scientists might have been employed to develop logical spaceship designs, the show's exact future date—and therefore its postulated technology—had not been established. In addition, scientific accuracy rarely yields saleable art. The design had to be esthetically agreeable to the mass audience; it also had to be practical for stage illumination and photography.

Fortunately, the production unit was blessed with several talented men who had aviation backgrounds. Walter "Matt" Jefferies, an antique airplane buff, was instrumental in gathering reference material and developing the spaceship design. After final approval from the producer, Jefferies had whipped up a prototype model by late summer of 1964.

The Howard Anderson Company, an optical effects house near the Desilu Studios, was contracted. Working with Jefferies, they constructed a tiny 4-inch wood-and-cardboard miniature. After approval what might be called the "publicity *Enterprise*"—a detailed 3-foot balsa-and-plastic model—was completed.

It was obvious that a larger miniature wired with lights would be required; the enormous cost would be quite a gamble at this stage, before a script had even been approved. Nonetheless, Anderson got the go-ahead to begin. Working with hardwood, balsa, plastic, and metal, the largest miniature ever made for a pilot film was constructed. Power for its numerous lights and flashers was wired through the pipe support of the model's base. The finished model measured 134 inches long; the saucer hull was 60 inches in diameter.

The *Enterprise* miniatures came almost four years before the science fiction spaceship world would be forever changed by the complicated "hardware" ships of *2001: A Space Odyssey*. Still, the *Enterprise* design remains a striking combination of logical prediction and hot-rod art direction. An example is seen in the "NCC" prefix to the hull number. Prior to 1950, standard American aircraft carried an "NC" license number prefix. Racing planes used "NR" and experimental craft were designated "NX." The prefix system could have been used to identify subsequent types of spaceships, since the *Enterprise* was at first a "cruiser class" vessel, neither the smallest nor the largest in what could have been a diverse fleet of star craft.

Test shooting got under way in September 1964. Although white coloring was tried, both models were eventually covered in a silver-gray coat. (Variable film stock, optical quality, lighting, and color television

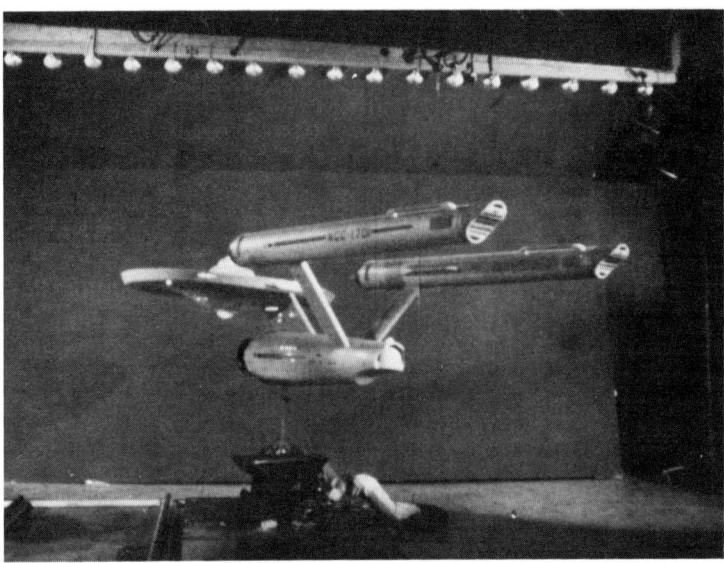
The 11-foot miniature shown shortly after its completion in the fall of 1964. Note the nacelle end caps and port side details, which were later modified.

Front view of the pilot film version shows even finish, enlarged sensor screen, raised bridge cap, and lack of detailing on saucer.

Pilot model nacelle details: Compared to the film production version, note smaller end cap with spike and lack of cross-hatching on the inside of the nacelle and support struts.

After the March 1966 improvements were made, only the model's starboard side was filmed. Note the globe-shaped end caps added to the nacelles, the cross-hatched panels added to the nacelle slots and pylons, and detail given to hangar bay.

transmission would make the ship appear white, green, and blue over the coming years.)

Pilot shooting was completed just before Christmas. Although the final edition of the 63-minute film remains a science fiction classic and television milestone, its professional appearance was hampered by the use of the two smaller miniatures. As is well known, a poorly conducted audience test led to its rejection and NBC's eventual agreement to a second pilot.

Although the theoretical size of the ship was enlarged for the second pilot, no major changes were made in the miniatures in 1965. The small cardboard-and-balsa model was no longer used and has since been lost. There was no money for an actual shuttle vehicle, so no hangar bay was necessary. In fact, publicity drawings made later that year showed flame and smoke blasting out of the lower aft end.

The new pilot, stalled from network viewing until the proper time in 1966, sold the series. Almost immediately the yard-long "publicity *Enterprise*" was employed in taking publicity pictures. By March of that year, the decision was made to improve the miniatures; models made for a chance pilot film could not hope to stand up to weekly viewer scrutiny. The following month saw the Anderson Company begin modifications to the 11-foot model.

The top saucer bridge section was lowered. Crosshatched panels were added to the nacelle struts and inboard nacelle slots. The "main sensor screen" was shrunk in diameter. Lighted pods were added to the nacelle "exhaust" cones. And, at last, the hangar bay doors at the aft of the secondary fuselage were defined.

The most difficult part of the project was the addition of the nacelle cap lighting effect. After removing

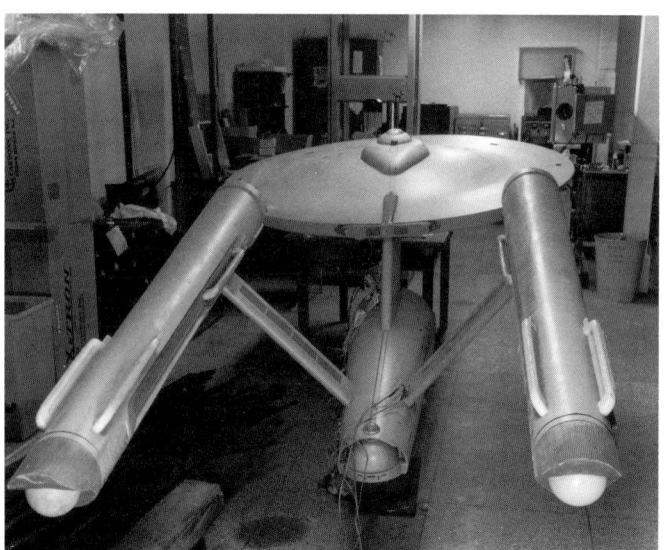

The Enterprise 11-foot miniature as it appeared at the Smithsonian Institution's Silver Hill, Maryland, restoration facility after being uncrated and reassembled for inspection. Exposed wiring on port side reveals why that side never appeared on the TV screen. Besides painting and other cosmetic work, the Smithsonian restoration team had to replace some mixing plexiglass windows, fabricate new front end caps for the nacelles, and install a new main sensor dish.

The 11-foot model was carefully weathered in 1966. Front view shows more detailing and smaller sensor screen.

The underside saucer numbers were turned around late in the first season to be more readable from the most used camera angle. The larger, spikeless cap replaced the nacelle cap used on the pilot film model.

and discarding the original spiked caps, the wood was hollowed out to house large motors. Small Christmas tree lights were added to newly fabricated disks, which were mounted on bearings and connected to motors for rotation. Slip rings were fashioned to provide power to the little winking lights; once the new translucent caps were added, no single light could be followed through the complete circle of rotation.

The heavy wiring required to operate the new effects could not be fished through the existing support pipe, so the entire port side of the miniature was torn up to route the cables. When the model was refinished with a beautiful "weathered" appearance, the port side was left unpainted; it would never be seen on camera again.

The smaller model, reclaimed from picture taking, was reworked to copy as many of the static changes as possible. The model wound up with a portion of its mounting bracket permanently affixed to the lower hull.

Publicity photos were taken of the 11-foot model on a darkened stage, allowing stars to be printed in later without a costly optical effect. For some questionable reason, publicity photos also were taken of the yard-long model suspended in front of a sky-textured background. These unflattering pictures, complete with visible wires and the dangling mounting bracket, have somehow become the most widely published photographs of the spaceship.

Since there was now no way to film the port side of the major model, and there was little reason to re-shoot a good library of stock footage, film of the older version was kept handy. While the smaller model was used occasionally in the series ("Tomorrow is Yesterday," "Requiem for Methuselah"), it was used mainly for publicity pictures with actors.

Little stock footage of its original version was employed; however, both versions of the larger model found their way into every one of the following 78

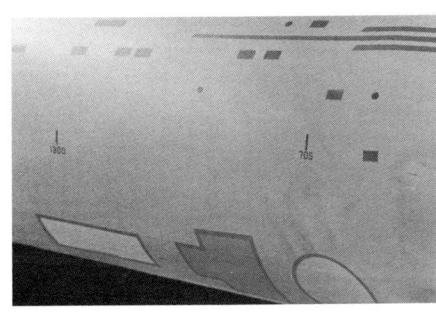

Detail in the form of markings adds to the visual excitement of the secondary hull. The windows are gray and black, the stripes are red, and the underside markings are various colors (the circle is yellow outlined in red; the T is gray outlined in dark gray; the rectangle is white outlined in black; and the square is white outlined in red).

The Enterprise as it hangs in the Smithsonian today. Blinking on-off lights in the new red nacelle front end caps have replaced the "pulsating" warp drives lights of the television series.

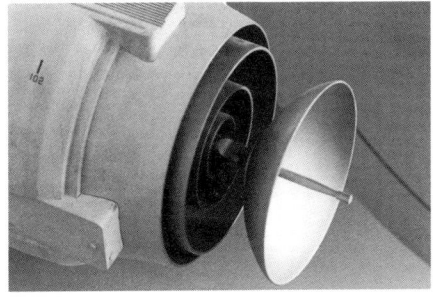

Close-up view of the main sensor dish on the secondary hull, an item that was lost or destroyed after filming of the new TV series, and had to be fabricated new during restoration. The dish is painted copper color.

episodes. Perhaps the confusion that could be expected from these circumstances is at fault, but many of the drawings since published of the *Enterprise* contain inaccuracies.

With the exception of a tiny metal starship trinket created for the first segment of the second season, no more professional *Enterprise* miniatures were made.

The second year did see several AMT model kits assembled to represent other starships in the episodes "The Doomsday Machine" and "The Ultimate Computer." Only the most minor changes, like turning the underside saucer numbers around 180 degrees, were made in the large miniature for the rest of the show's life.

The producer managed to salvage the yard-long model when the show was canceled in 1969. Shoved into storage, the major *Enterprise* decayed until Paramount donated it to the Smithsonian Institution. Restored by the addition of new nacelle caps and a new main sensor screen, the *Enterprise* is now on display in the National Air and Space Museum.

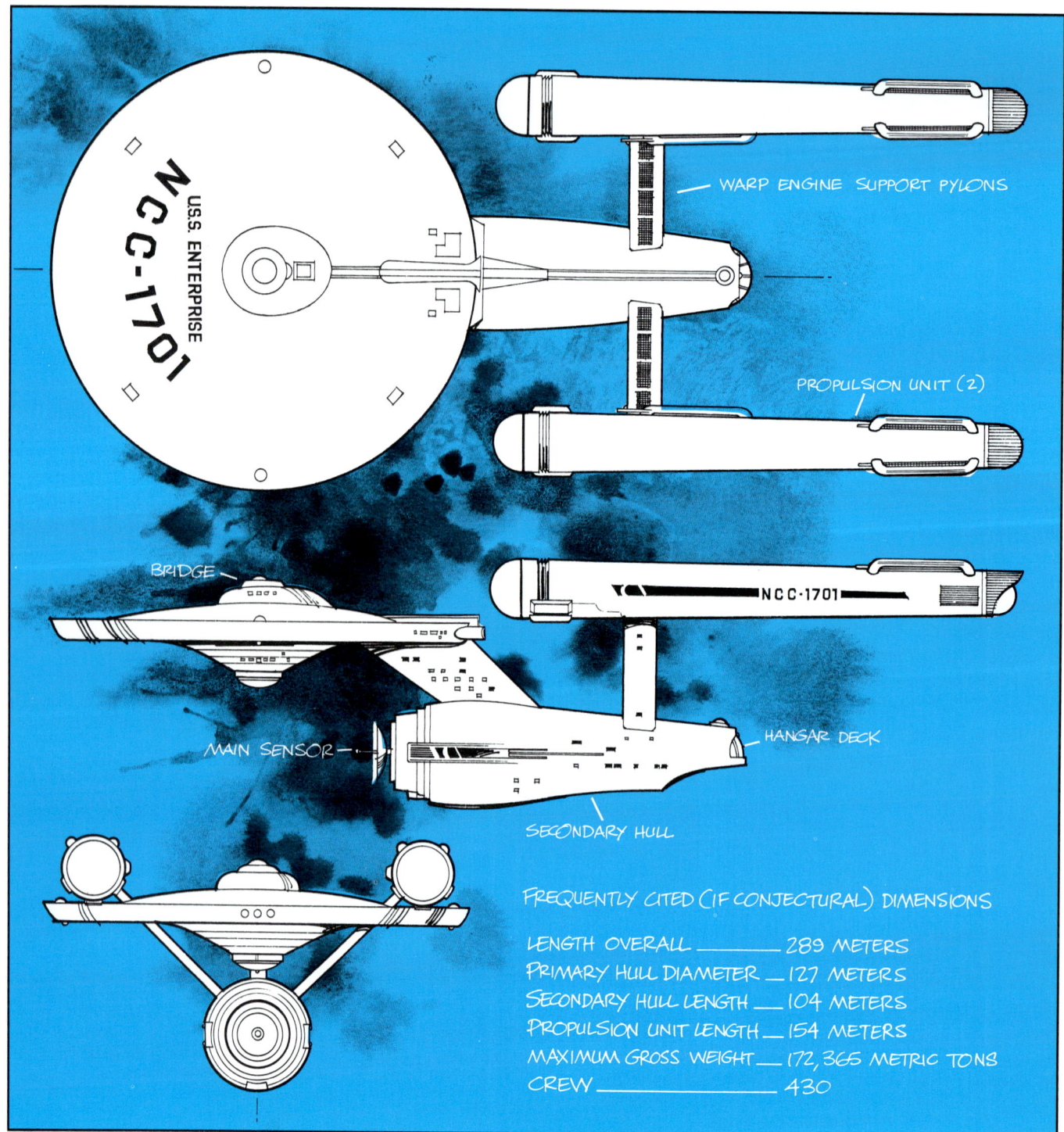

WEATHERING IS KEY TO X-WING

DON KLEIN

Kit: Luke Skywalker's X-wing fighter from *Star Wars*
Manufacturer: ERTL/AMT, U.S.A., Kit No. 8918FO
Scale: Unstated; approximately 1/48
Dimensions: Length—11⅞"; height—2⁹⁄₁₆" (closed wing configuration, gear down); wingspan—10¹⁹⁄₃₂"; vertical wingspan—4¼" (open or attack wing configuration)

ERTL/AMT's reissue of the MPC rebel X-wing fighter from *Star Wars* is not as good a kit as the firm's TIE ship, but it is a fundamentally sound model with quality modeling work. With care, it builds into an excellent replica of the battered "good guys" ships in the movie.

I concentrated my efforts in three areas while building two of the X-wing fighters. First, I set out to solve a bad mismatch problem between the top and bottom halves of the main fuselage section. Next, I added a number of minor details to the body and wings to make the model more interesting and more authentic. Finally, I chose to execute the weathering on one ship—the version with the wings in the takeoff/landing configuration and the gear down—with powdered pastel chalks, but to age the second model with airbrushed paint.

Correcting the fuselage mismatch. To start on the body mismatch problem, cement the fuselage top and bottom

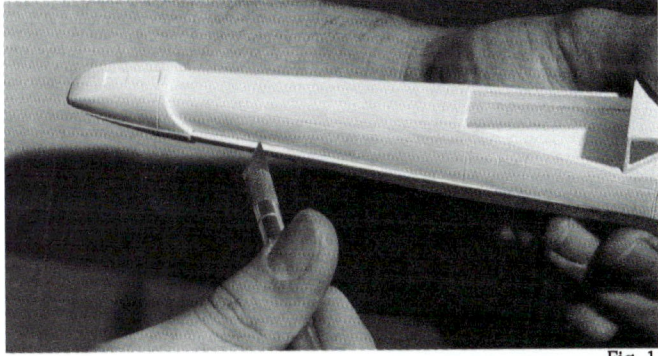
Fig. 1

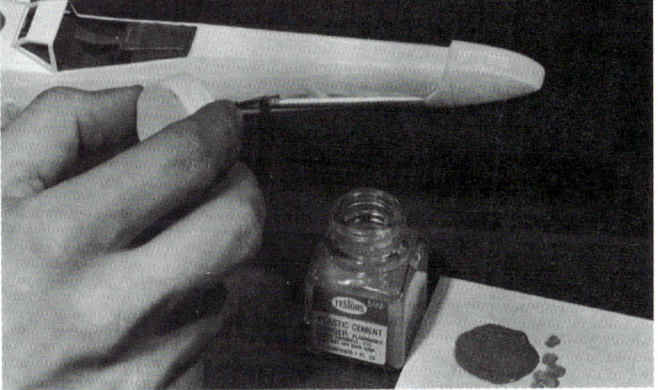
Fig. 2

together, carefully positioning the top half so the overlap is equal on both sides. Make sure this bond is a solid one, and give it at least overnight (two days isn't too long) to dry.

I filled the joints between the fuselage halves with green automotive spot putty thinned with a few drops of Testor's (liquid) Plastic Cement No. 3602 to make it easier to work. Check the photos and note the shelf-like "step" (Figs. 1 and 2) where the bottom fuselage juts out wider than the top section from the cockpit forward to the nose cowling. This is intended, so don't try to fill it in. Use Flex-I-Grit sanding film, wet, to smooth your filled areas.

As usual, I filled in one or two of the panel grooves inadvertently while correcting the fuselage mismatch and had to rescribe the joint lines. Do the recutting with a No. 11 X-acto blade. Where the length of panel line to be scribed is short, you can get away with slow freehand cutting; but when the line exceeds half an inch or so, use a small steel straightedge or other piece of metal as a guide to ensure that the blade doesn't wander.

The kit instructions show how to assemble the pivoting X-wings inside the body so they will move from the takeoff/landing position to the attack position. This is fine for a quick-and-dirty toylike model; but for a scale model, forsake the operating feature and leave the instruction assembly sequence behind as well.

It would be almost impossible to properly work on the main body joints with the wings in the way, so I cut the wings apart—making them four separate wing sections. This allows you to add the wings after the rest of the model has been assembled, painted, and decaled, and to add extra detail to the model wings, should you choose to assemble them in the attack configuration. You trade the operating feature (of dubious value anyhow) for better access to the parts during construction and easier handling all around.

The landing gear parts on my kits were poorly molded—the old problem of mold misalignment on round parts. This kit is far from unique in this respect—it's a problem that seems to plague the majority of plastic kits. If you're looking for the ultimate in appearance, the solution is to fabricate replacement struts from plastic or brass tubing and stretched sprue, but I made do with the kit parts by carefully scraping the mold parting lines with a modeler's knife and finishing up with files to achieve a correct round cross section.

Adding some simple details. The final modifications I made to the kit were adding several dozen small bits of styrene sheet to the body panels as extra relief detail, and rebuilding the laser cannons with stretched-sprue tips. There is nothing exacting about the detail—it's just small rectangles of .015", .020", and .025" styrene sheet cemented in place to represent access hatches, protruding junction boxes, and so on. On the attack model, I detailed each recessed area on the wings with bits of styrene sheet, sprue, and Plastruct T-section stock.

After the cockpit interiors were painted, the clear areas of the canopy were masked and the fuselage and wings airbrushed with the base color—Humbrol underside white for the landed model, a 3:1 mix of Pactra white and Pactra aero blue on the craft in the attack configuration. With the base coat cured and stable—at least two days' drying time—I added the decals, coated them with Dullcote, and started the weathering treatments by adding black ink in the cracks between the panels to accentuate them. I did this by applying Pelikan brand Plaka ink No. 70, using a fine brush. As you apply the ink, use a bit of rag moistened with water to rub away any excess. Rub toward the rear of the model so that any resulting smudge will simulate the residue of oil or fuel leakage.

Weathering with pastels. Weathering helps to blend any model into a more homogeneous whole by covering decal edges and small imperfections with a grubby film. It also enhances realism, of course, and brings out detail.

Inexpensive pastels are available both as individual sticks and inexpensive sets at most art-supply stores.

(Top) Plaka ink was used to accentuate paneling. (Bottom) Chalks were used for weathering. Note several shades of grey paint on engines.

X-wing—and use them for practice before moving on to areas where less dirt is needed.

The finishing step for a chalk weathering job is to make sure it stays put. Use a clear fixative to hold the powder in place and prevent fingerprints from showing. Since brushing on a clear coating would disturb the chalk particles, use a spray can or airbrush to add the fixative. This can be an artist's matte fixative specifically intended for pastels (beware—it may attack the decals), or a clear flat model coating such as Testor's Dullcote. Be sure to mask clear canopy areas before applying the clear coating.

Weathering with the airbrush. My attack configuration model was weathered with an airbrush, but its paint job before weathering differs considerably from the landed version. First, I brush-painted several fuselage panels a slightly contrasting gray to highlight them. To indicate sheet-metal areas that had been replaced but not repainted, I omitted four pieces of the broad red fuselage decals and painted two panels just aft of the nose cowling a silver color to represent the new metalwork. I also dabbed small amounts of the base color over the decal stripes around the edges to simulate peeling red paint. I then added ink to the panel grooves as on the other X-wing.

Before starting with the airbrush, I masked the cockpit windows and a couple of panels that I wanted to look new, and removed the R2D2 unit. I did not weather the R2D2 unit on either X-wing model, because leaving it shiny points up the fact that it is a drop-in part of the craft, and reminds the knowledgeable viewer of the details of the movie plot.

The colors I used for airbrush weathering were Pactra tinting black, Pactra white, and Floquil roof brown (very little) mixed in various combinations as I went along. Mix them in a ratio of approximately 1 part paint to 1 part thinner.

Adjust the paint feed and air supply on your airbrush to make

Start by grinding small amounts of dark brown, rust brown, light gray, and black chalk into little piles of powder by rubbing them on a coarse file or on a rectangle of medium-grit sandpaper. Don't grind much—you'll find that a little goes a long way.

Now dip your brush into one of the powdered chalk piles and dab the powder onto the model. After you first touch the brush to the model, streak the powder toward the rear, and tail off the streak to give it a fading effect. A No. 2 or 3 round sable brush is a good size for manipulating chalk. Work the powder into cracks and around the edges of raised detail, and always finish with a rearward motion. If you overdo the amount of chalk, simply brush it out. Proceed cautiously until you get the hang of the process. Start on the areas of the model that would probably be the dirtiest—the engines, on the

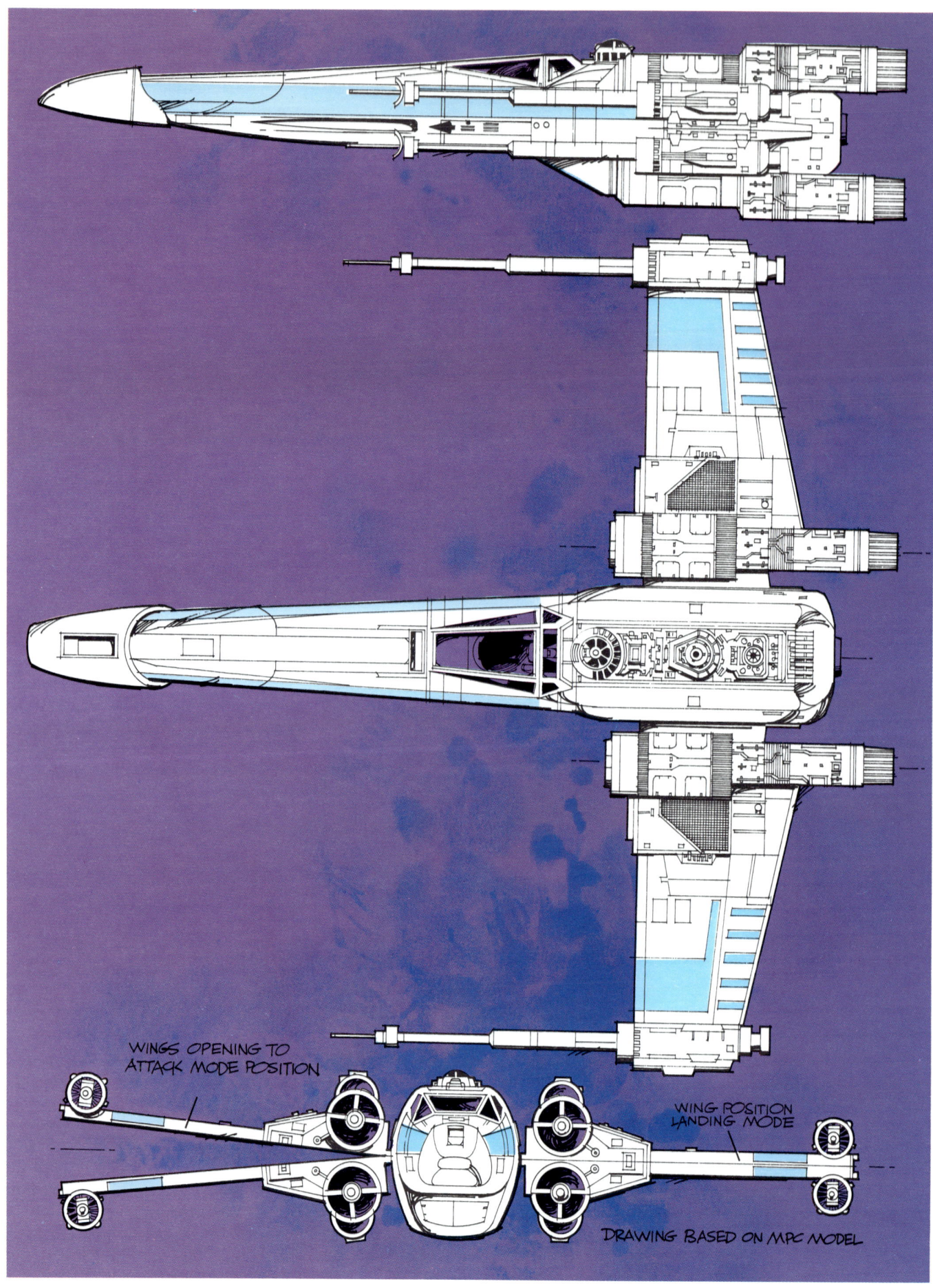

fine—about 3/64" wide—lines, and leave them at this setting. Apply paint to the model in much the same manner as the chalk on the earlier version, always streaking the color toward the rear and allowing it to tail off. In some areas I just shot short bursts of paint onto the model. With the airbrush aimed toward the rear, use a lighter weathering treatment on the insides of the wings; these portions of the craft would not normally be exposed in atmospheric flight.

After I was satisfied with the basic weathering treatment, I went over the model and added a few highlights with pastel powder, especially around the engine nacelles. These small touches don't require clear fixative spray unless the model will be handled frequently.

More notes. The poorest moldings in the MPC X-wing kit are the pilot seat, the pilot, and the little R2D2 droid. I improved the seats on both models by adding rounded cushions made from 1/16" styrene sheet. Although my model with the gear down excludes the pilot, the figure in the attack-configured craft is a jet fighter pilot from a Tamiya 1/48-scale A-10 kit. I used standard military-miniature painting techniques to highlight and shade the figure to bring out the detail.

To rework the R2D2 unit, first sand away all the raised detail lines on the dome-shaped upper portion; fill any bad joint lines or mismatches with putty; and wet-sand to a smooth, regular hemisphere. You need not bother with the bottom portion of the unit, since it will not show on the finished model. Add a new ring of .004" sheet plastic at the point where the rounded part of the droid joins the cylindrical portion, and then dress up the dome with rectangular and square bits of sheet plastic and small slices of stretched sprue. I finished by airbrushing the unit silver. The five metallic blue segments on the dome were painted by hand. As I noted before, the R2D2 units are purposely not weathered.

My models show but two of the myriad variations possible on this attractive fighter. Here are a few more suggestions: an open canopy with a truly superdetailed cockpit interior, a battle-damaged version with evidence of laser hits on the engine nacelle and the R2D2 (as in the movie), or even a diorama crash scene. The rest is up to you.

MASKING AND PAINTING THE TIE

DON KLEIN

Kit: Darth Vader's TIE fighter from *Star Wars*
Manufacturer: ERTL/AMT, U.S.A., Kit No. 8916FO
Scale: Unstated; approximately 1/24
Dimensions: Wing length—6 13/16"; Fuselage length—5 3/4"; wingspan—7 7/8"

The batlike Twin Ion Engine (TIE) fighter was the better of MPC's two *Star Wars* spacecraft (now reissued by ERTL/AMT). The fit of all major sections is excellent, and the detail is intricate and well-executed throughout. The recessed, cut-in panel separation lines—which, in terms of making the molds, is the more expensive way to do the job—are superior in all respects to raised panel lines. These greatly help the realistic appearance of the completed model, and they facilitate panel masking for painting the model in several subtly different gray shades.

The areas on which I concentrated with this modeling job were panel masking and interior detailing. The main objective of the panel masking and painting was to make this model interesting and bring out its detail without a weathering treatment. I took on the cockpit detail as a challenge, mostly because the large scale of the model lends itself to such detailing without making the detail gross or clumsy. The cockpit can be completed as a subassembly and slipped whole into the model, another factor which makes the job easy.

Masking and painting. After carefully examining the model, removing the parts from the molding trees, and cleaning away all wisps of flash and mold parting lines, I tested the major parts for fit. I then airbrushed

 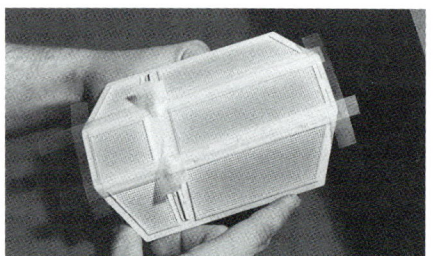 

Solar panels dried two days before masking with transparent tape, then were airbrushed flat black.

all parts with a slightly grayed white—in this case, Humbrol underside white mixed with a few drops of Pactra tinting black.

After giving this coat a couple of days to dry completely, I masked the areas between the solar panels on the angled "wing" structures with Scotch Brand Magic Transparent Tape. This masking was easy because the solar panels had straight edges that could be masked with tape right off the roll. I then airbrushed the waffle-texture solar panels with flat black.

Masking the multi-shaded panels on the main body section of the TIE fighter is more complex, but not necessarily more difficult. I started by laying strips of the Scotch transparent tape over the entire fuselage. When the fuselage was covered, I used a sharp No. 11 X-acto knife blade to cut through the tape around all the panels that I wanted to airbrush in contrasting shades of gray. I used the fine grooves between panels as guides, placing the knife point in them and lightly slicing through the tape. I then pulled the tape off the panels, airbrushed the appropriate color, let it dry, and remasked before peeling off another set of panels for the next color.

The sequence should be:
(1) cover entire fuselage with transparent tape
(2) cut through on all panel lines
(3) peel away tape to expose panels that will be one color and airbrush
(4) remask and recut panel lines
(5) peel away next color and paint again.

For small raised details, I used a brush to paint the contrasting grays. The grays I used were Pactra rebel gray and Pactra panzer gray.

This is a good place to stop and add a word or two about applying and removing masking, whatever kind you use. The idea of any masking, of course, is to separate colors cleanly without damaging the paint underneath the masking medium.

Here are some tips to help you accomplish that. First, off-the-roll masking tape has a fuzzy, indistinct edge. It's certainly good enough for full-scale subjects such as automobiles and motorcycles and house windows, but the fuzzy edge is alarmingly apparent in our scale models. You can improve on it somewhat by using Scotch Magic Transparent Tape, which has a much sharper edge. Or you can apply your masking tape to a sheet of glass and cut a sharper line with a single-edge razor blade or hobby knife held against a steel straightedge. This will give you a cleaner edge to the tape and cleaner separation lines between colors.

Applying your tape to glass first also removes some of the stickiness from the tape adhesive, and this is good too. Most tapes are, if anything, too sticky and will pull paint.

When you apply the tape, press it down only firmly enough to make a good seal along the separation line, and press down only along the line. The idea is to achieve the minimum bond between tape and paint needed to prevent the second color from getting under the tape. The less force used to bond the tape to the paint, the less likely the tape will pull the paint beneath it when the masking is removed.

Where large areas away from the separation line must be masked from overspray, use newspaper instead of masking tape. Use tape along the separation line, of course, but slip a sheet of clean newspaper under the opposite edge of the tape and use it to wrap the model.

Leave masking tape on the model for as little time as possible. This means you should mask a model, paint the second color in the same work session, and remove the masking as soon as the second color is set—not dry, just set. If the masking is only on the model for two hours, that's good—if only for one hour, that's better. The longer the masking remains on the model, the more it bonds to the paint coat beneath it, and the more liable it is to pull the paint when it is removed. So do your masking, paint, and remove the masking. Also, make sure the paint coat that you mask over is cured. This may mean waiting two days or even a week between colors, but the alternative is a ruined model.

When you remove tape, peel it off by pulling the end of the tape directly back over itself. This will mean that the tape is at close to a 180-degree angle where it peels off the model. This extreme angle places the stress of tape removal on a very small surface area of paint. This, in turn, minimizes the total adhesive pull on the paint, which can be a lot if you pull the tape straight away from the surface. It's similar to trying to hold up a heavy object with tape—you can't do it with a narrow band; but if you use wide tape or several narrow strips, the total adhesive power will do the job. We want to minimize the holding power here.

Getting back to the TIE fighter, after completing the major panel masking and painting, I used a 000 spotting brush to pick out details such as holes, sockets, and fastening devices with flat black and dark gray.

Other than the cockpit detailing, the only modification I made to the model was to omit the landing gear parts. Landing gear is in the optional category on any model, and I omitted it here to emphasize the ship's horizontal lines.

Cockpit detailing. The cockpit parts in the kit provide good molded-in detail but lack the three-dimensional look. The biggest drawback is the lack of a cockpit seat. I omitted the Darth Vader figure, so my cockpit detailing involved scratchbuilding a pilot's seat from sheet plastic and adding control levers, instrument dials, indicator lights, and hand grabs from various bits and pieces of stretched sprue.

The seat is the first order of business. I made mine from .025" sheet styrene and added cushions made from 1/16"-thick stock. The edges of the cushions are rounded, and their glossy red finish simulates fabric. I made handles on either side of the seat and added dials and indicator lights made from stretched sprue.

Scratchbuilt seats, dials, and indicator lights bring color to cockpit interior.

The dials and indicator lights in the remainder of the cockpit were made from bits of stretched sprue, too. The trick is to slice off bits of sprue until you have some thin round slices. You may have to make a dozen slices to get three good dials or indicator lights, but the raw material is cheap.

After painting the dials and indicator lights in various colors, I made them stand out in the flat interior by touching them with a drop of Devcon "5-minute" Epoxy applied with a toothpick. Each drop of epoxy dried in a transparent bubble shape, giving dimension to the features on the instrument panel.

The detailing is freelance, because the available drawings don't offer much in the way of detail for this portion of the ship. The idea is to make the cockpit look functional and interesting—how you do it is your own business, so have a ball!

I also picked out the molded-in cockpit detail with various gloss and transparent paints.

I did not weather the TIE ship, other than adding a few small streaks around what represents operating machinery. Why? Well, this ship is the personal fighter of Dark Lord of the Sith Darth Vader, and it's supposedly the most advanced of its type. So it's new, and it probably gets tender loving care, sort of the general's personal jeep. Also, it doesn't make any atmospheric flights as the X-wings do, so it isn't subjected to the ravages of wind, dust, and water.

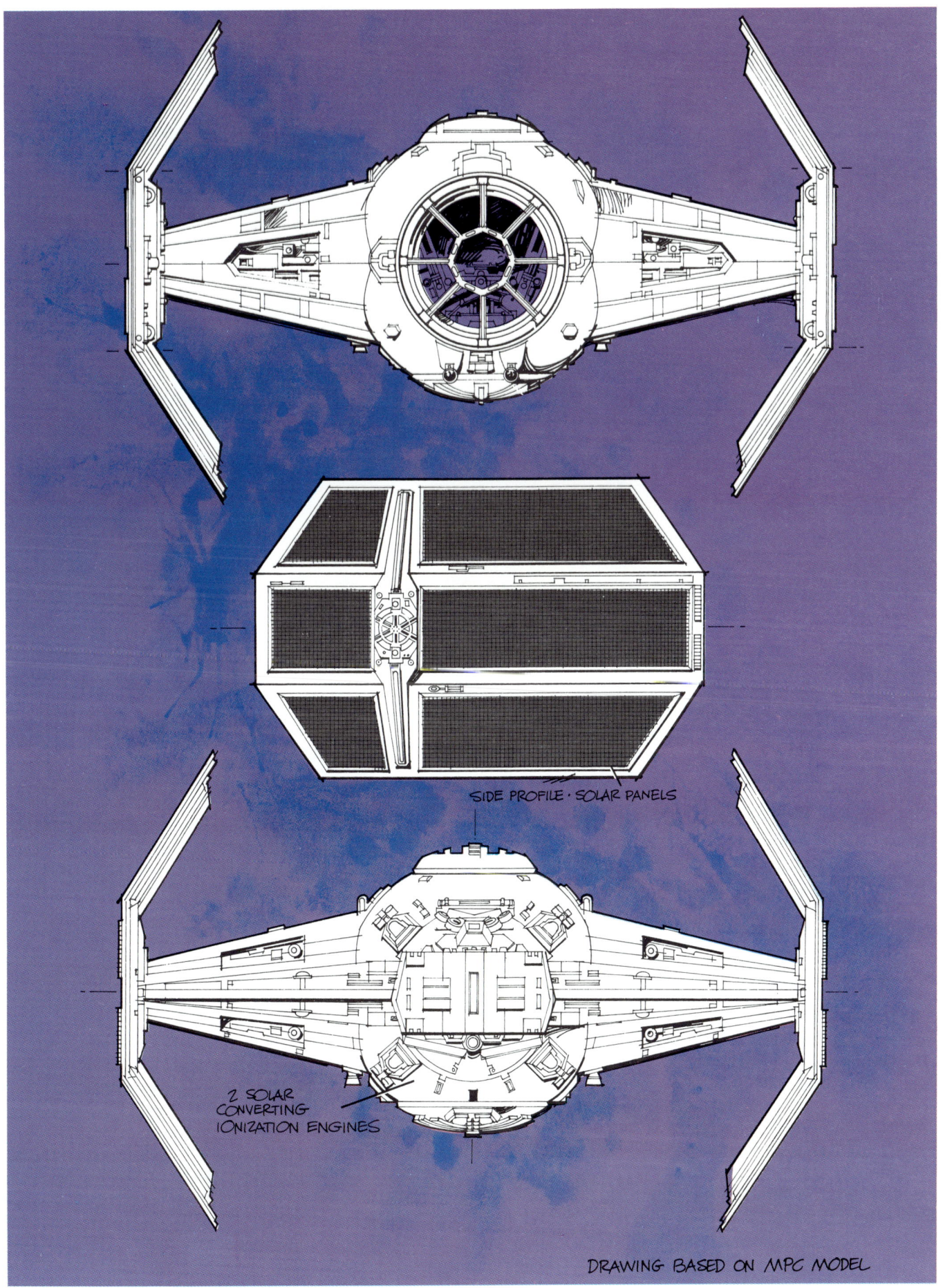
SIDE PROFILE · SOLAR PANELS

2 SOLAR CONVERTING IONIZATION ENGINES

DRAWING BASED ON MPC MODEL

VIPER AND RAIDER

Shapes of good and evil from TV's Battlestar Galactica

CHRIS TIETZ

Editor's note: This and the accompanying two chapters on the Viper and Raider kits were first published in the original edition of this book. At this printing, these kits had been out of production for some years. Interest in them, however, remains sufficient that we felt their inclusion in the second edition was called for. You may still be able to find one of these kits in a hobby shop, or at a swap meet or modeling show. If you are on-line, you may also search a scale modeling newsgroup for information. And finally, scratch-builders will find the information here vital to their projects.

Battlestar Galactica fills the television screen with action-packed derring-do among the distant stars, exposing audiences to new and exciting alien civilizations and technologies. Glen A. Larson is Executive Producer and creator of the series; Universal Television is the studio; and ABC-TV (which telecast the first story in the fall of 1978) is the network. The story revolves around a race of peaceful humanoids who are sneakily attacked by a rival race of armor-clad mechanical beings known as Cylons.

The peaceful race, which exists in 12 colonies on 12 planets, has unwillingly battled the Cylons for a thousand years. Weary of war, its Council of Twelve has finally negotiated a tentative peace agreement with the Cylon Empire. As they trustingly approach the Cylons in deep space with most of their combined worlds' spaceships, including enormous "Battlestars," the peace-seeking humanoids are betrayed, attacked, and virtually wiped out, as are their home planets. Only the *Galactica* ship, temporarily diverted from the main fleet, escapes destruction. Horrified, the *Galactica* crew helplessly observes the holocaust via long-range viewers.

When the *Galactica* returns to its home base, the outraged crew finds a burned-out world. Gathering together what's left of their belongings, that world's survivors begin a shaky Dunkirk-like exodus in a rag-tag fleet of spaceships, using any and all types of spaceworthy vehicles. These tattered remnants rendezvous with *Galactica,* and together embark on a long voyage in search of the legendary planet

Earth, where they hope they can establish roots on a new globe. This immense journey is punctuated by encounters with alien beings and hazardous space phenomena, and is underlined with anxiety over being caught by the ever-pursuing, always vicious Cylons.

The *Galactica*. The *Galactica* itself is a 2,000-foot behemoth of a spaceship with a strikingly reptilian appearance. Its wedge-shaped bow cuts a domineering swath through space. Two huge rectangular pods straddle the center of its body in gigantic outrigger fashion. The twin pods serve as hangars for the ship's shuttlecraft and its fleet of attack-intercepter Viper ships.

The Viper intercepter. These one-man intercepters are 29 feet long, with three fuelburning rocket engines that have inner atmospheric flight-and-landing capabilities. They achieve extreme acceleration in space by use of turbo thrusters, and can throw themselves into an immediate reverse direction: a useful ability in battle situations. Twin laser guns are mounted along each side of the Viper's fuselage, just below the cockpit, jutting forward from the wing fronts. The hangar pods of the mother ship *Galactica* provide complete storage and maintenance facilities for the Vipers. A Viper is shot into space through one of 20 launch tubes lining the outward side of each pod. To re-enter a hangar, the small intercepters approach the rear of a pod and plunge through the landing bay door's gaping mouth—which is also the entrance/exit port for shuttlecraft.

The Cylon attack ship. Ever appearing to menace our space travelers on their odyssey are the manta ray-like Cylon ships, with rounded wing shapes that curve down at the ends as if to encircle and engulf their prey. These inner/outer atmospheric, twin-engined, three-robot ships have twin-mounted laser guns, one on each side of their louvered cockpit shields, and are 35 feet wide by 26 feet long. The Cylon ships harbor themselves aboard their floating headquarters, which is shaped like two huge saucers joined, and separated, by a thick central column. The entire configuration spins ominously through space in search of the fleeing humanoids, ever ready to launch its fleet of attack ships against *Galactica*.

The 2,000-foot *Galactica* is actually a 6-foot-long model equipped with air-cooled quartz-light engines

and dozens of tiny, glowing optical fibers that serve as viewports. Its surface is painstakingly detailed with cannibalized model-kit parts. Only one model exists. By means of multiple exposures shot from many different angles, it represents an entire fleet.

The Viper intercepter ships, built in 1:16 scale, are manufactured from a mold that produces six basic components forming one complete ship. Such ease of assembly permits rapid construction when models are needed. No explosion versions exist. Since all detonations are effected optically, models are never lost. Three of the Viper models are specially equipped with nitrogen "engines." Liquid nitrogen is injected into the airflow that passes around hot quartz lights, producing a showy vapor trail to indicate that turbo thrusters are operating. One Viper model is fitted with landing gear.

Like the Vipers, the bat-type Cylon ships are mold-made and feature air-cooled quartz-light engine effects. These models are built in 1:24 scale.

Lasergun effects for both types of ships are achieved optically. A full-size Viper intercepter ship is used in scenes with actors. Because of the greater dimensions of the Cylon ship—a result of upping the crew from one to three robots—only a full-size mock-up of its cockpit area was built.

DRAWINGS BASED ON MONOGRAM MODEL

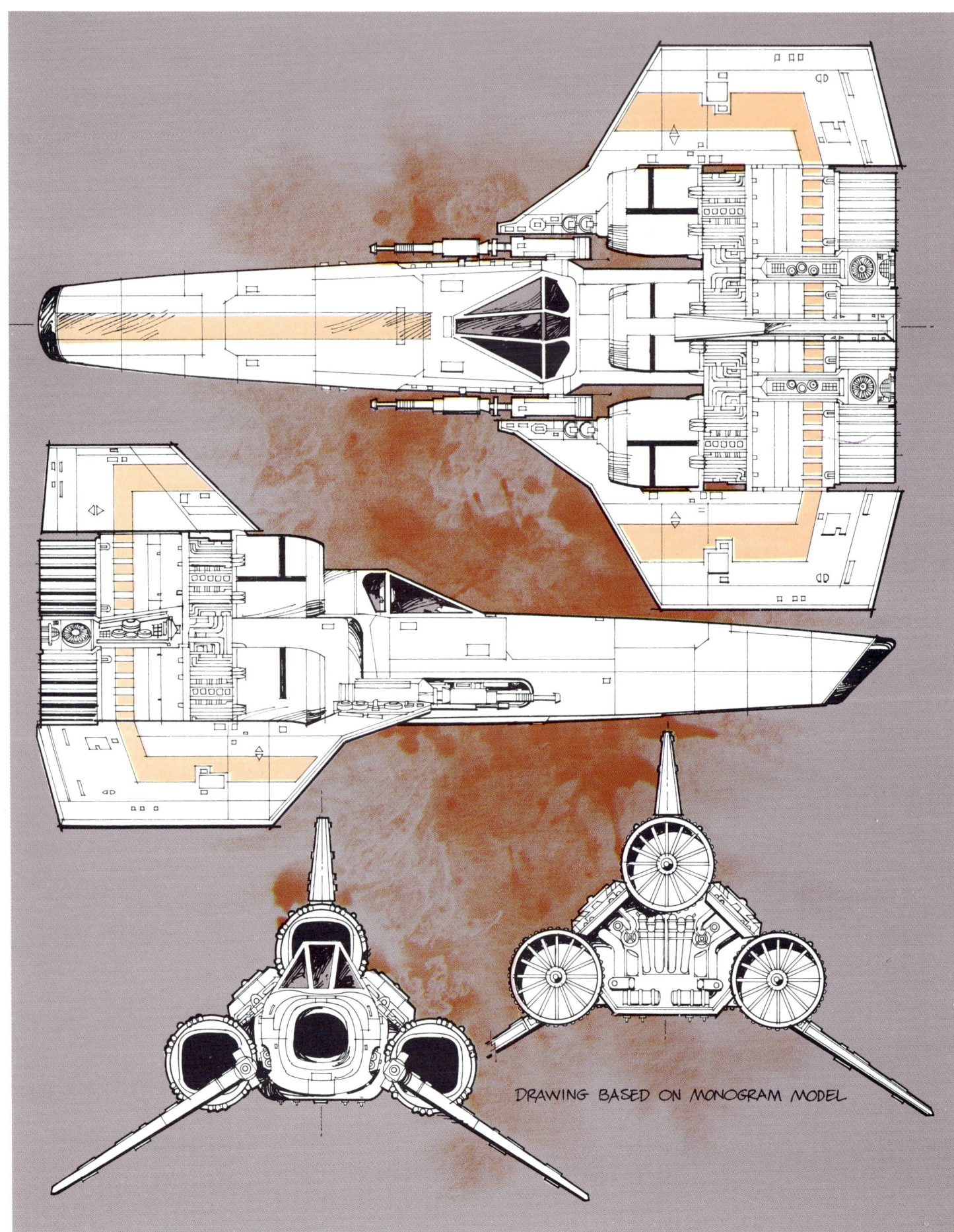

FIBER OPTICS LIGHT THE RAIDER

BOB HAYDEN

Kit: Space Fighter Raider
Manufacturer: Monogram, U.S.A., Kit No. 6026
Scale: Unstated: approximately 1/35
Dimensions: Length—8 13/16"; wingspan—10 11/16"; height—2 13/16"

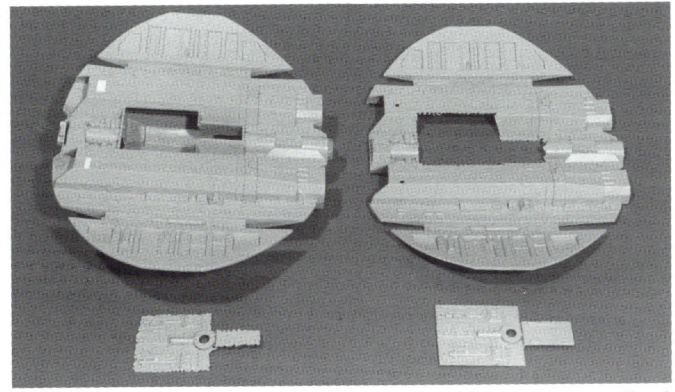

Fig. 1

Editor's note: This and the accompanying two chapters on the Viper and Raider kits were first published in the original edition of this book. At this printing, these kits had been out of production for some years. Interest in them, however, remains sufficient that we felt their inclusion in the second edition was called for. You may still be able to find one of these kits in a hobby shop, or at a swap meet or modeling show. If you are on-line, you may also search a scale modeling newsgroup for information. And finally, scratch-builders will find the information here vital to their projects.

Monogram's kit of the evil-looking Cylon Raider fighter from *Battlestar: Galactica* captures the basic forms and intricate detail of the original, but a major problem is the poor fit of the fuselage sections. These are very large injection-molded pieces—almost 11 x 8½"—and they don't mate well. In fact, both sections must be flexed considerably to get them to mate at all. The instructions, which are structured around the working missile launchers, depict the fuselage halves going

Fig. 2

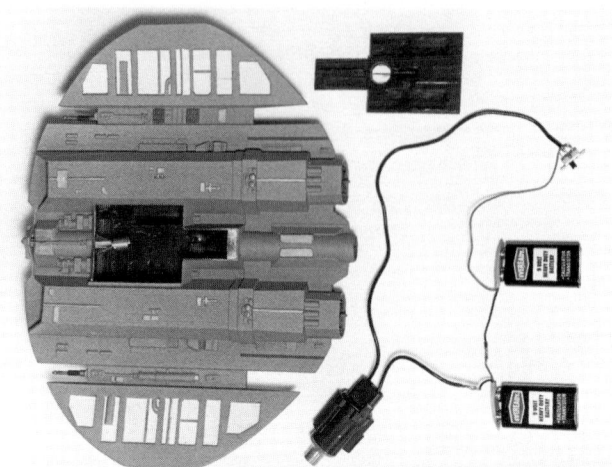
Fig. 3

together last. This sequence of assembly is almost unthinkable for scale model construction, because we want to fit, putty, and finish the major fuselage joints early on in the assembly process.

Because of the filling, filing, and sanding time required, I did not assemble the model with major modifications or detailing, but I did add a moderate amount of fiber-optic lighting.

In cutting an access opening for the fiber-optic lighting system, I substantially increased the flexibility of the bottom fuselage section. This, in turn, made mating the sections easier and also allowed join-line filling before the addition of the lighting system.

Making the access hatch. The slide-out opening that I devised is similar to many portable-radio and tape-player battery compartments. Building this feature required two Monogram kits: one to furnish the bottom section into which the opening was cut, and one to provide the hatch. Start work on one of the unassembled kit sections by drilling several dozen holes around the periphery of the "hatch" piece in the bottom (Fig. 1). Use a fine jeweler's saw to cut through the plastic between the holes, and remove the hatch. Carefully finish the edges with a medium-cut file.

Next, drill holes in the bottom section that will be used on the finished model. Stay inside the hatch area with your drilling and, again, connect the holes by sawing. Carefully file the edges of the opening smooth, checking the fit frequently with the hatch. Cement styrene strips in place to make retaining slides for the main section of the hatch, then fabricate and install the hold-down cam-tongue arrangement for the narrow rear portion of the hatch.

Discard the working rocket launchers entirely— they complicate assembly, the action feature is of little value, and the TV craft do not launch similar ordnance. Cover the front launcher holes with round plugs made from the tips of the missiles, and add short sections of styrene strip to cover the actuator holes on the bottom of the craft.

Fuselage assembly. Next, assemble the fuselage halves.

I bonded one area at a time. First, bond one side "wing" area, and allow overnight setting. Then, bond the central sections; again, allow overnight drying. Finally, bond the remaining "wing." This approach permits stressing of the parts into proper alignment while using a limited number of clothespins and spring-action clips.

You'll find many places where the fuselage halves don't come together in a smooth edge. Use a file to make the majority of these areas presentable, but resort to filler putty in some spots, notably around the front of the cockpit area and around the engine exhaust ports. The rear central fan housing also requires a built-up ring of putty to properly fair it into the rear of the fuselage.

After completing the putty work and wet-sanding all refinished areas with 400- and 600-grit paper, I coated all puttied areas with Floquil Barrier (Fig. 2). This prevents the solvent in the paint or primer coat from softening the surface of the filler putty. Often, a paint such as Floquil, even when applied almost "dry" by airbrush, will soften filler putty. When the putty dries, it will shrink just a little, opening cracks around its edges. The Barrier coat defeats this tendency.

At this point I added the rest of the kit parts. I also cut plugs from heavy sheet styrene and added them to the exhaust ports, then cemented the exhaust-port inserts—part 11— to this plug. This simple modification blocks the interior from view, making the model appear more substantial and realistic.

The lighting system. Before painting, I worked out the details of the lighting installation. Adding lighting to a plastic model, especially a small- or medium-size one, involves a tradeoff. On one hand, the finished effect is usually excellent, but on the other, there are many problems: added weight, access to the components, heat buildup inside the model, and light leaks. All these problems take time to solve, and you should realize that including even simple illumination effects will take a good deal longer than building a "straight" model.

I decided on fiber-optic lighting for the Raider model for several reasons. First, and most important, fiber-optic light guides allow you to use a single light

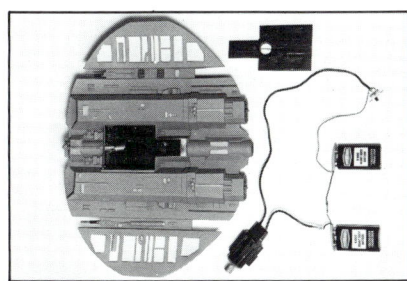

source—one bulb—to take light to more than one location. This cuts down on heat buildup inside the model, and it also means that there is only one bulb to burn out— and to replace. Finally, using only one bulb means that there is only one source of light leaks to seal off. The fiber-optic light guides and light source that I used are sold by The Hobby Factory, P. O. Box 67, Abington, PA 19001. I used their No. 40 hobbyist kit, which provided several sizes of light guides, a 12-volt focused light source, and colored dyes (Fig. 3).

Accomplishing as much pre-installation work for the lighting system as possible before doing any painting prevents the model from becoming worn and dog-eared from the handling during lighting installation. After deciding where I would place the individual lighting elements, I drilled holes for each and checked the fibers for fit. I then made a lens on the end of each fiber by holding a lighted match near it as described in the instructions. After once again checking for fit, I removed the fibers, labeled them as to location, and dyed the ends. I also assembled and checked the light source and batteries for fit inside the model.

With all holes drilled and the groundwork for the lighting installation complete, I airbrushed the inside of the model with flat black enamel to minimize reflections. I then airbrushed the exterior with a dark metallic gray made by mixing a small amount of Floquil platinum mist into stock reefer gray.

I applied Floquil Crystal-Cote on the areas where the decals were to go and then applied the decals. I spent one evening picking out raised panels and details with a slightly contrasting metallic gray flat enamel. I also used other colors— black, red, deep blue, and bright silver—to pick out raised details and add a three-dimensional quality to the model. The recessed panels on the underside of the craft were painted off-white.

When I was satisfied with the paint scheme, I airbrushed the entire model with Testor's Dullcote. This coating is available only in spray cans, so in order to airbrush it you have to discharge the can into a bottle! This may seem like a very uneconomical way to do the job, but the Dullcote is a very good matte finish and well worth the extra step and extra expense.

Final lighting installation. Adding the lighting completes the job. I wired the light source to two 9-volt transistor radio batteries in series, giving a theoretical 18 volts. The actual voltage is more on the order of 14 volts, which will not blow out the automobile bulb in the light source. I modified the light source by cutting back the conical end housing and epoxying a short piece of 3/8" o.d. brass tubing in place. This makes it possible to gather together and illuminate a larger bundle of light guides. I found that the fiber bundle I made would fit into a piece of 1/4" o.d. tubing, so I used three short sections of the intermediate tubing sizes to bush the bundle up to a slip-fit into the tubing section on the light source. These telescoping tubing sizes—available at almost every hobby shop—are handy for many projects. In this case, they allow removal of the light source from the model for bulb replacement but maintain good alignment of the light source and light guides while in the model.

Touch up the sides of the light guides where they are external to the hull of the ship. Then drop in the light source and batteries, push the switch to "on," and turn out the room lights to get the full effect of the lighted miniature.

BATTLE DAMAGE ADDS REALISM TO VIPER

BOB HAYDEN

Kit: Space Fighter Viper
Manufacturer: Monogram, U.S.A., Kit no. 6027
Scale: Unstated, approximately 1/32
Dimensions: Length—11"; wingspan—5 13/16"; height—3 15/16"

Editor's note: This and the accompanying two chapters on the Viper and Raider kits were first published in the original edition of this book. At this printing, these kits had been out of production for some years. Interest in them, however, remains sufficient that we felt their inclusion in the second edition was called for. You may still be able to find one of these kits in a hobby shop, or at a swap meet or modeling show. If you are on-line, you may also search a scale modeling newsgroup for information. And finally, scratch-builders will find the information here vital to their projects.

Realism, not perfection, is what most of us strive for in scale model-building. Many modelers weather their models to add realistic evidence of use and age and to bring out detail. Battle damage on combatant models is part of weathering, and I built Monogram's excellent Viper kit specifically to try out some battle-damage techniques.

I built two Vipers. The first was built "stock" to acquaint myself with the kit, to identify any problem areas, and to determine the best assembly sequence for the damaged version. Although it certainly takes longer than just diving in, I recommend this "stock kit first" pro-

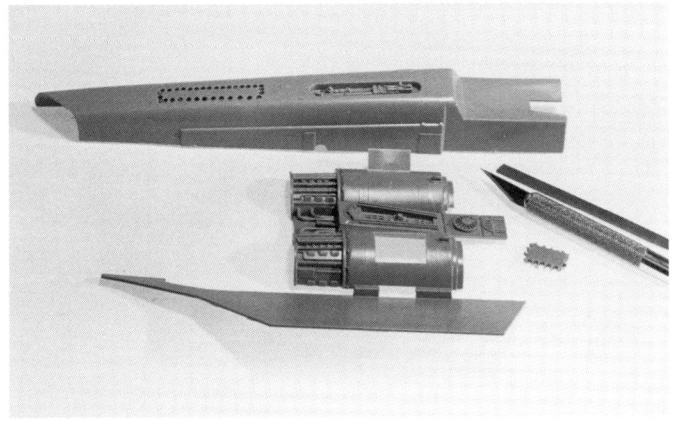

Fig. 1

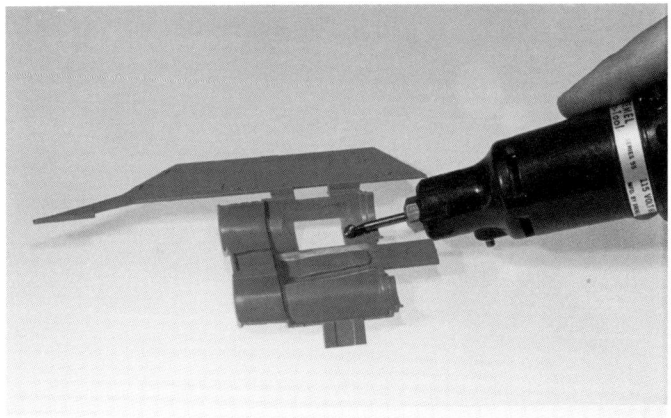

Fig. 2

cedure any time you plan to modify a kit extensively. Before building either kit, I planned the damage I wanted to add. I found an easy way to record my ideas was to attach a sheet of tracing paper over the photo on the box top and doodle away. This way, you make your mistakes on paper, where they are easy to erase, and check your ideas against the unassembled model parts.

I decided to add only moderate battle damage to my model to represent a craft that had sustained two or three close laser-cannon hits, but that was still flyable. I also wanted to build the damaged version using only one kit (often, extensive modifications require a second kit to provide "fit-in" panels or hatches where holes are made in the first kit). I settled for the damage shown in the photos: a peeled-back sheet-metal panel on the left side of the fuselage near the nose, a blown-away access-hatch cover on the number two engine, and a skin rupture on the port wing.

Both models share the same basic paint scheme. The overall color is a light gray made by adding small amounts of silver to white. This gives you a color that has more "life" than a gray made by adding black to white. The fuselage nose fairing and the rear engine casings are a dark metallic gray, also with some silver mixed in. The striping was done using the kit decals.

Let's get to the heart of the job—the battle damage treatment. Bear in mind the battle-damage techniques can also be used to add details such as open servicing hatches to undamaged models.

The first task was to remove the forward fuselage panel and the "blown away" hatch on the engine housing. Many modelers favor a jeweler's saw for this kind of cutting, but I find I can work faster by drilling many holes around the perimeter of the area to be removed. I used a 3/32" drill and left a 1/64" safety margin between the holes and the edge of the opening (Fig. 1). After drilling the holes, cut through the narrow bits of plastic connecting them, remove the waste material, and finish the edges of the opening with files. Work to the line slowly and carefully.

The full thickness of the plastic is much too great to simulate sheet metal, so the next job is to remove material from the back of the parts to make the exposed edges thin. I used a steel cutter In a motor tool to do this work (Fig. 2). A slow-speed control is a must here—if the tool runs fast, it will melt the plastic instead of cutting it. I used a ball-shaped cutter. Here again, work slowly and carefully, removing only a little material at a time.

The peeled-back fuselage panel is made from .010" styrene sheet. Cut a strip to the correct length and width to fit into the opening in the fuselage. Then, hold the strip over the barrel of a hot knife (a soldering iron will do the same job) to soften it. While the plastic is slightly rubbery, deform it as shown in the photos (Fig. 3). Next, cement it in place with Micro-Weld and smear 5-minute epoxy over the joint on the inside for strength.

To make the impacted area on the wing, I slowly warmed—again, using the hot knife—an area on the underside of the top wing section until it sagged downward. I allowed this sunken area to cool and used the ball cutter in the motor tool to cut through the plastic from beneath. The ragged hole that this makes does a good job of representing a glancing laser cannon impact.

Before we can start assembly, we need to settle on the inner structure that will show through the damaged openings. The small hole in the wing is the easiest. I backed the hole with a small scrap of corrugated aluminum siding—the kind sold for model railroad structures—and added three sizes of wire to represent cables and hydraulic lines (Fig. 4). I painted this green to contrast with the wing color.

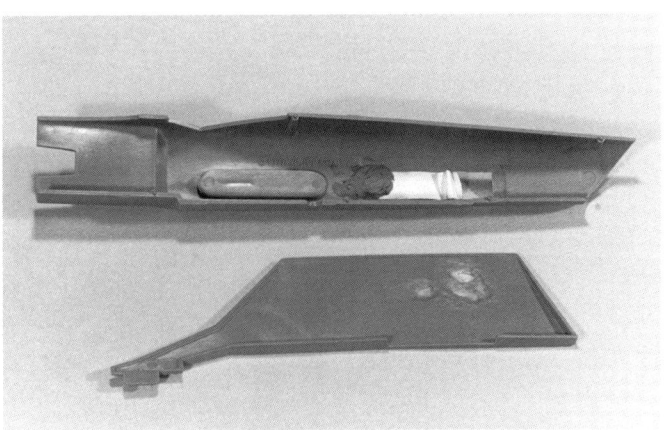

Fig. 3

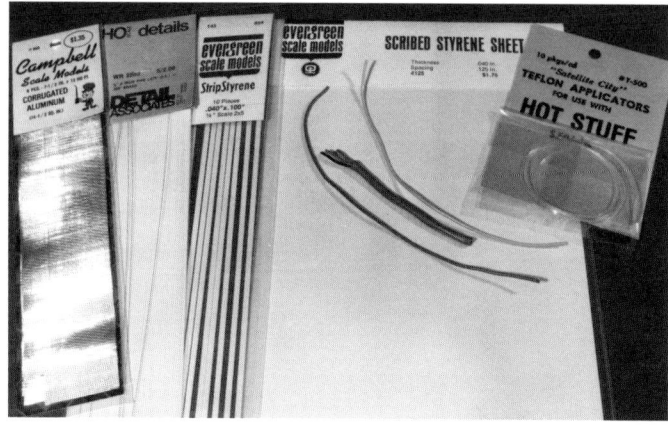

Fig. 4

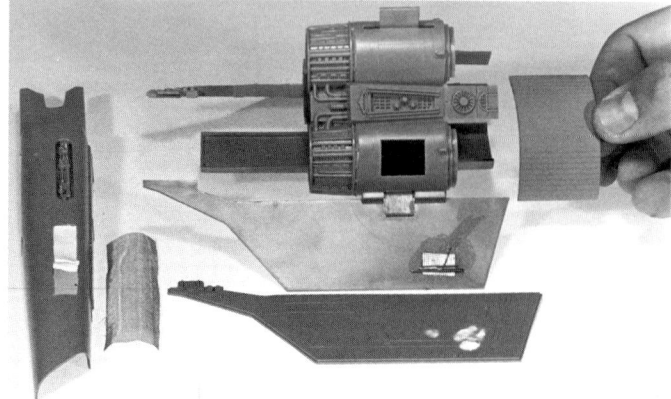

Fig. 5

Fig. 6

The peeled panel in the fuselage is backed with 70-mesh brass screen painted light blue. To this I glued several sizes of small wire and two pieces of small-diameter stranded insulated wire to represent a ruptured cable. A few small rectangular styrene pieces represent electrical junction boxes and instrument sensors.

To back the blown-away engine service hatch, I cut a piece of scribed styrene to fit inside the engine housing and pre-curved it by rolling it around a ¾" dowel (Fig. 5). Next, I painted it orange and added another stranded-wire ruptured cable, two clear fuel lines (made from the Teflon tubing that comes with Hot Stuff brand ACC), a piece of wire, and three junction boxes.

I then removed the inserts and sprayed the inside of the fuselage flat black to ensure that any light that gets into the model will not reveal the lack of complete detail. I assembled, painted, and decaled the model using standard kitbuilding procedures (Fig. 6), maintaining openings through which I could position the battle-damage inserts. I added the inserts during final assembly.

One item that is lacking in the kit is a black decal piece for the small cockpit window pane on the top surface of the cockpit. I made one by cutting the correct shape out of a scrap decal. To keep the window panes shiny while spraying the rest of the model with clear flat finish, I cut individual masks from Scotch Magic Transparent Tape and lightly pressed them over each window area. After applying the flat finish and pastel-chalk weathering, I removed the tape, revealing the glossy areas.

Other modifications that come to mind for the Viper fighter include replacing the opaque cockpit windows with clear plastic, adding cockpit detail, and adding ski-like landing skids. The best bet for landing gear would be to utilize strut parts from an aircraft model and make the skids from sprue.

Ready for launch, a spacecraft is inspected by a robot in the hangar bay of a mothership.

SCIENCE-FICTION MODELING WITH JUNK

Tips for finding detail treasures among your discards

BRIAN TREMBLAY
MODELS BY DEREK NADEAU

Have you thrown out that old hair dryer? Did you sell that Polaroid instant camera at your garage sale? Have your kids worn out their GI Joe or *Star Wars* toys and accessories? You may be surprised at the detail that can be gleaned from unwanted items like these. Don't throw anything away before examining it closely. Obsolete cameras are excellent sources for springs, lenses, and other intricately contoured parts.

Ping-Pong balls become fuel tanks, or when cut in half, the basis for gun turrets. Oxygen or fuel tanks can be made from empty CO_2 cartridges for pellet pistols. Flexible drinking straws make good fuel lines and conduits—even exhaust pipes. Grilles can be fashioned from mosquito netting or window screening.

Don't overlook clear plastic blister packs—they may find just the right contour for your scratchbuilt spacecraft hull. Strengthen the flimsy plastic by smearing 5-minute epoxy on the inside.

Basic details. One of the most useful items for detailing is a set of old bass guitar strings. You probably have a youngster in the neighborhood who is (or knows someone) in a rock band. Just ask for the old set of strings. By removing the hard wire core, you can flex the outer winding to suit your needs. They make excellent conduit and piping.

To mask the origins of your details, airbrush a light coat of paint and weather with pastel chalks.

Pictures of computer screens in electronic hardware catalogs can look like working on-board computer systems. If the back of the illustration isn't printed, you could make it look even better by backlighting with a small, low-power bulb from a hobby shop.

Tonal technique. To achieve different hues on hull surfaces, try this technique. First, add a "grain" on a panel by sanding it in one direction, then sand adjacent panels in different directions. The grain of the panels will reflect light at different angles and the surface will appear to be made of many separate pieces. More tones can be produced with paint. Add a few drops of white, black, silver, gray, blue, and brown to small amounts of "steel" enamel. Painting individual panels with different colors suggests different metal alloys.

Combining the sanding and painting techniques produces panels that appear to change color and reflectance when the model is viewed from different angles.

Other useful items:
- Plastruct structural shapes such as I beams and channels
- Old window screens
- Seats from old model cars
- Telescoping radio antennas
- Grilles from broken hair dryers
- Insulated household wire

So watch carefully what you toss into the trash can. "Junk" modeling can be exciting, fun, and full of surprises.

Cockpits of spacecraft can be enhanced with seats from your old, discarded car models. Check out blister packaging for preformed plastic shapes that can be used for hulls and cockpit transparencies.

The back end of this spacecraft is highly detailed. Look carefully and you'll see CO_2 pellet gun cylinders, Ping-Pong balls, and old bass guitar strings.

Check for futuristic-looking weapons in GI Joe and *Star Wars* toy sets.

Luke and faithful R2D2 are speeding through the Death Star trench, but Darth Vader, in TIE fighter, has them in his sights.

THE SHIPS OF STAR WARS

CHRIS TIETZ AND GEORGE ELRICK

In 1977 *Star Wars* detonated in terrestrial movie houses with the impact of an exploding supernova, increasing science fiction fandom a hundredfold. Its overwhelming appeal has been attributed to several self-evident facts: (1) It creates a dazzling new mythology; (2) It simplifies life, depicting an extreme "us versus them" polarization of good combating evil; (3) It parades seemingly endless special effects before audiences who have hungered for them since Stanley Kubrick's *2001: A Space Odyssey*; and, (4) It introduces seven interstellar vehicles the likes of which have never been seen on the screen: realistic spaceships, some of which impart a nitty-gritty aura of having been drop-kicked around a football field.

The story itself is fairly simple: "A long time ago, in a galaxy far, far away," a bunch of good guys heroically tangles with a bunch of bad guys. The bad guys are Imperials of the evil "Empire," which has spread, Hitler-like, from planet to planet and from solar system to solar system. The good guys are the rebels: a tattered Alliance which aims to restore the honor and integrity of the Old Republic, even if this entails perishing in electronic shoot-outs that make the gunfight at OK Corral seem like patty-cake. From their secret hiding place on one of the planet Yavin's moons, the rebels pick at and sting the Imperials with everything at their disposal, including an outnumbered fleet of one-man fighters that are somehow reminiscent of the plucky R.A.F. soaring up to butt heads against Nazi bombers.

What catapults the plot from the commonplace to the sublime is the genius of writer/director George Lucas, who demanded a "well-used" universe for his film. No matter how extravagant their derring-do, the heroes always seem to have dirt on their shoes and a

Full-scale Y-wing and X-wing rebel fighters are serviced before the battle.

Battered and peeling, Han Solo's trilobite-shaped *Millennium Falcon* has nevertheless eluded many Imperial patrols.

threadbare look about their clothes. All well and good. What really captivates audiences (modelers in particular) is the revolutionary appearance of the spacecraft. Totally absent are the smooth-skinned saucer contours and streamlined fountain-pen shapes normally linked with science fiction films. What do we have in their place? Believable spaceships, many of which have about as much cosmetic appeal as a secondhand car.

In contrast to the pampered vehicles of the Imperial fleet (the bad guys), all ships belonging to the rebels (the good guys) look weathered, beat up, and worn. They're discolored with dirt and grime from endless planetary landings and take-offs. Their burnt and scarred hull plates testify to scores of previous battles. They unabashedly display areas blackened by engine exhaust. Much of their piping is exposed, as are many of their engine parts. They're magnificently angular, practical, and functional, projecting a scorn for mere appearance. Their design not only reflects the reel world of space, but NASA's no-nonsense know-how with the real world of space.

The opening chase scene of the rebel blockade runner trying to flee a much larger enemy ship sets the pace for the entire film, immediately establishes an "us versus them" syndrome, and introduces the first two mind-blowing spaceship configurations.

The Rebel Blockade Runner. Princess Leia Organa's starship suggests a dimly remembered past when spaceships were designed with wedding-cake flair and character. This outdated vehicle—wearing its age with grace and dignity—has an elongated body replete with interesting shapes and angles. It was originally intended

Rebel X-wing fighters, using tactics similar to those of World War II aerial combat, peel out of formation for attack on Death Star as Y-wings follow

to be Han Solo's pirate ship but was soon deemed too pedestrian for that role, as well as too similar to *Space: 1999*'s *Eagle*. Therefore, it's used as a "throwaway" for the initial hare-pursued-by-the-hounds sequence in which it's inexorably overtaken and lifted, via force beams, into the bowels of an Imperial Star Destroyer.

The Imperial Star Destroyer. This wedge-shaped leviathan of space used by the evil Empire is thousands of feet long and as sinister in appearance as a cobra flicking its forked tongue in and out. When it first soars overhead, filling the entire screen with its massive presence, it elicits oohs and ahs from the intimidated audience. Imperial Star Destroyers ceaselessly patrol the *Star Wars* galaxy, relentlessly enforcing the imposed dictatorship.

One of those opposing the Empire is a maverick young space smuggler named Han Solo, who inadvertently lets himself become a pivotal figure in the rebel revolution, though he plays this role, at first, purely for mercenary reasons. Han is a reluctant hero who comes through when the chips are down, aided by Chewbacca, the shaggy Wookie. His souped-up Corellian freighter—the *Millennium Falcon*—is one of the most fascinating ships in the film.

The Millennium Falcon. This oversize, one-of-a-kind "hot rodder" makes supralight jumps into hyperspace whenever pursued by authorities: a capability not without peril since accurate navigation at such velocities is difficult. A second-rate pilot might find himself slamming through the flaming heart of a star, crashing into a suddenly looming planet, or disintegrating into subatomic nothingness in a black hole. But Solo is an unexampled pilot, and his ship is a first-rate craft.

Though it's fast, the *Falcon* is no beauty. Loading mandibles protrude from its bow, and its stern is crammed with a doctored-up propulsion system. For the purposes of attack and defense, its topside and

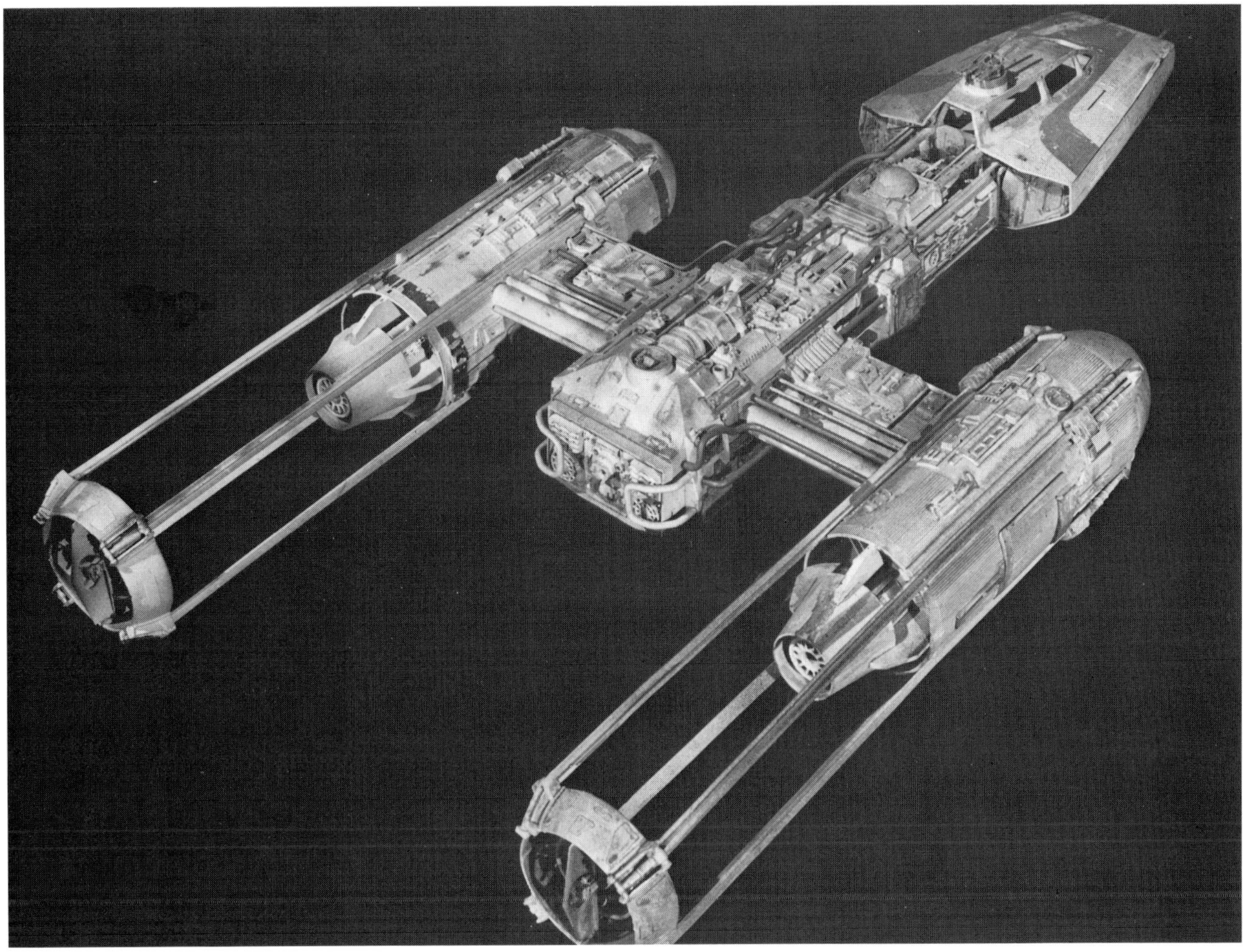

Y-wing nose represented streamlining that once extended length of craft, but was discarded to ease maintenance. Model is light gray with yellow stripes.

bottom-mounted turrets are equipped with laser cannons. The unlikely looking spacecraft plays a key role in *Star Wars,* carrying the rebel protagonists in and out of Imperial entanglements, and ultimately coming to the aid of Luke Skywalker as he destroys the giant Death Star in the climactic battle.

Just as the *Falcon* may be somewhat shabby, so is the rebel fleet of X-wing and Y-wing fighter craft. As mentioned in the first paragraph, each looks as though it has been drop-kicked around a football field—and, in a sense, it has. In observing the screen's slam-bang encounters, one surmises these one-man ships resemble poor relations because they see an incredible amount of day-in-day-out action, which causes endless, hurried repair work. Devoted mechanics—who faithfully keep moving parts, propulsion systems, and weaponry in well-oiled shape—have time only to splice bare replacement panels into the skinwork without worrying about such niceties as applying touchup paint.

The mechanics' ceaseless efforts are rewarded by the fact that these formidable fighters can blast skyward from underground hangars in a split second and can go 15 rounds with anything their size or bigger. A testimony to the once-large size of the rebel fleet—now

worn down by attrition—is the extensive use of cannibalized parts for repair work. Let's examine these mighty mites in more detail.

X-wing fighter. Each X-wing has a slender, 29-foot body terminating in a quad of fuel-burning rocket engines for space travel, plus air intakes for atmospheric flight. Its wingspread spans 25 feet. The horizontal wings, which provide lift within the atmospheric envelope, separate into an "X" ("guns at ready") position once the fighter

97

enters the near-vacuum of space. The X-wing has four wing-tip laser cannons for maximum firepower plus proton torpedoes, which are blasted from tubes flanking the fuselage and—in the film's climactic struggle—are used to destroy the Death Star battle station.

The X-wing is a one-man ship only in the sense that a solitary human occupies the cockpit. But that human isn't really alone. Squatting in an external aperture behind the cockpit is a highly programmed R2 droid unit, whose function is to serve as an onboard mechanic and troubleshooter. It may have a metallic exterior instead of a soft skin, and complicated electronic circuitry rather than nerves and blood vessels, but its "fix everything" presence enables the pilot to focus his attention and reflexes exclusively on fighting.

For firing the lasers, the X-wing pilot has avionic help in the form of a half-manual, half-automatic gunsight with electronic crosshairs. For firing the proton torpedoes, the pilot has fully electronic computerized targeting in a pull-over head-up display (although Luke Skywalker eschews the computer for the more accurate guidance of the "force" in releasing the proton torpedo that destroys the Death Star). Defensive equipment includes a warning scope on the control panel and built-in deflector shields that can be angled in any direction.

The offensive armament of the X-wings—in particular the proton torpedoes and targeting computer—suggests that the fighters were designed as much for attacking fixed or planetary targets as they were for engaging other ships in air-to-air or space-to-space combat. The heavily weathered appearance, in addition to suggesting frequent use combined with the capability of maintaining only priority items, also indicates a preponderance of atmospheric missions (on which weathering would be extreme), perhaps against planetary occupational forces of Imperial storm troopers.

Finally, an irony concerning the X-wings: They are not referred to by that name at any time in the movie (neither are the Y-wing nor TIE fighters mentioned by name). The name grew out of informal use at Industrial Light and Magic, the firm which filmed the special effects. By the time the novel appeared that was derived from the screenplay, the name X-wing had caught on and was used prominently; in fact, the X-wings at one point are given an even more formal designation: Incom T-65.

Y-wing fighter. The Y-wings are fewer in number than the X-wings and are not as fully developed as to their capabilities. The name is derived from their configuration or shape. In the film, they make the first, ill-fated attack down the Death Star's trench, with results somewhat reminiscent of Torpedo Squadron 8 at the Battle of Midway.

The climactic battle scene that eventually—and shatteringly—wipes out the awesome Death Star is depicted with different details in the book than in the

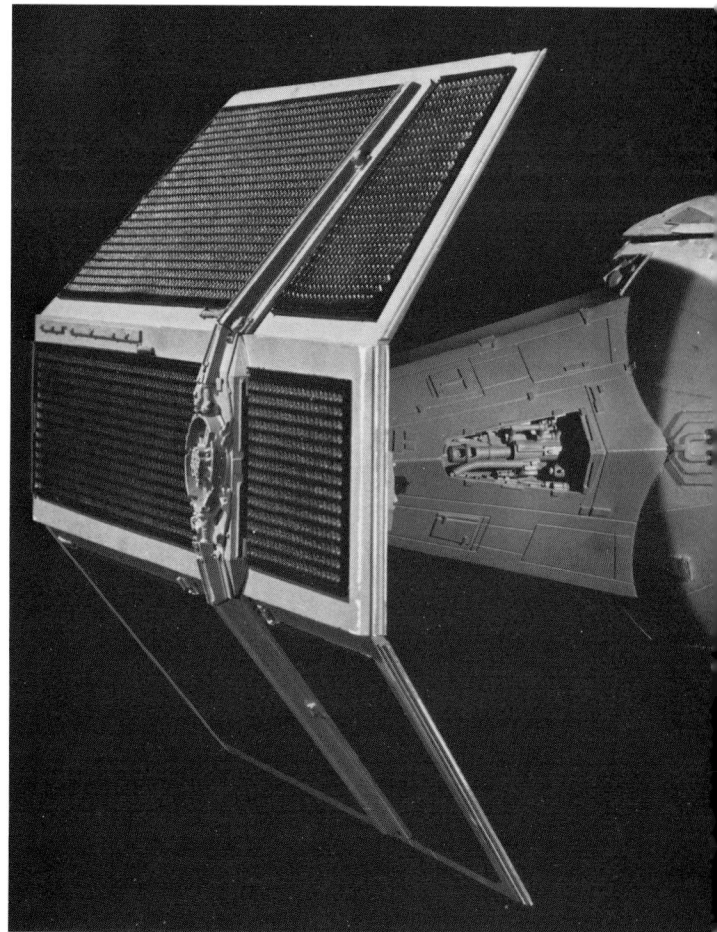

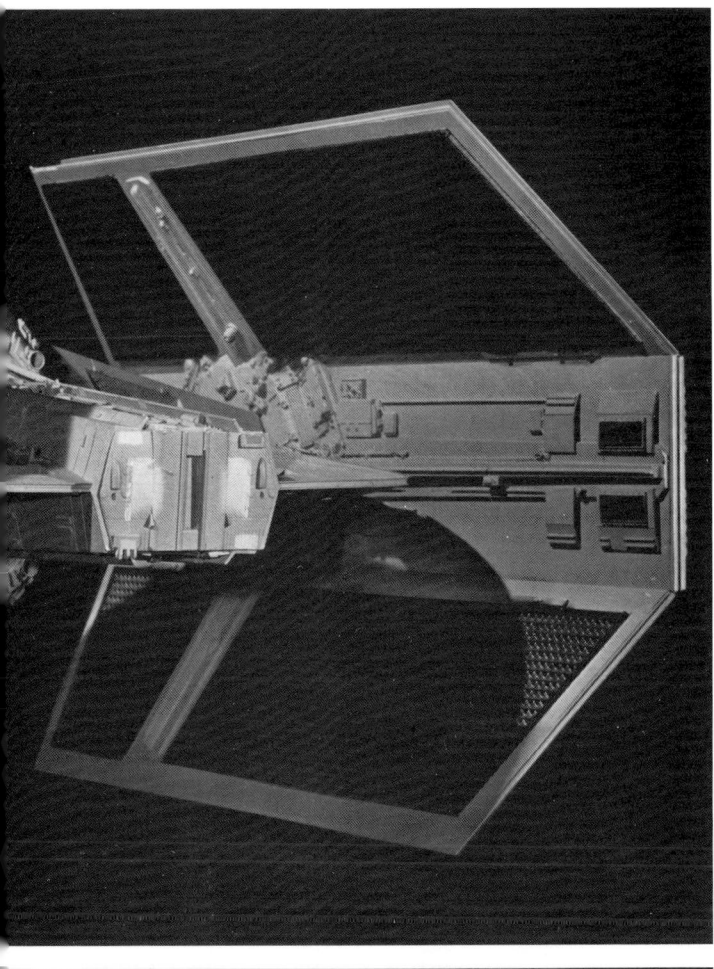

In order to eliminate unwanted reflections during filming, no glass was used on canopies of studio models of the two versions of the TIE fighter. Laser cannons distinguish "top" from "bottom."

movie. In both, the rebel attack fleet is launched from a junglelike moon of the gaseous planet Yavin. In the book, the attacking fleet is identified by the Imperial defenders as containing at least 30 ships. Four squadrons, in fact, are launched, with a mix of X-wing and Y-wing fighters. In the initial formations, Yellow Squadron covers for Red on the first run, and Green covers Blue on the second pass; after that it's pretty much every man for himself. Luke Skywalker's ship is "Blue Five" and has blue markings.

During the prolonged attack, individual rebel pilots identify themselves, via their helmet transceivers, as "Yellow One," "Red Three," "Green Four," and so on. In the film, Luke's ship is "Red Five" and has five red markings on its wings. Only X-wings from Red Squadron are shown, although ostensibly other X-wing squadrons take part in the battle. The few Y-wings shown are part of Gold Squadron. In the movie, Imperial defenders count 30 ships comprising the attacking force. In both book and film, most of the attacking fleet is wiped out—some by the Death Star's defensive guns but most by defending TIE fighters led by Darth Vader.

As the Death Star atomizes in a holocaust to end all holocausts, the novel depicts several X-wings and Y-wings, plus the *Millennium Falcon* (which made a ninth-hour appearance to disrupt Darth Vader's pursuit of Luke Skywalker's deadly torpedo run), making their return to the rebel base.

Before the curtain is yanked down by Luke, the rebel fleet dogfights with scores of deadly TIE fighters—the insidious Empire's counterpart of X-wings and Y-wings. Let's see what makes the TIE so fearsome.

Imperial TIE (Twin Ion Engine) Fighter. Unlike X-wings and Y-wings, TIE short-range, one-man fighters are basically employed in space's semi-vacuum, not within planetary atmospheres. Again, unlike the motley-looking X-wings and Y-wings, they're store-window clean, indicative of the fact that they're transported hither and yon in the massive bellies of Star Destroyers, from which they're launched to do sentry and reconnaissance duty, to patrol the Imperial forces' outer domains, to provide defensive protection for the all-important Death Star, and to summarily slap down any stirrings of revolt.

An individual TIE is menacing in appearance, with two laser cannons mounted just below its spherical cockpit's circular viewport. Support columns stretch out from each side of this bubble, firmly clasping two oversize, hexagonal, solar-energy-gathering sails which transmute photons into the considerable quantities of electricity needed to operate the double ion engines. In the two engines, positively charged atoms are accelerated at tremendous speeds by electrostatic fields, then neutralized by regaining their electrons a split-second before being ejected as an enormously forceful propellant through a rear nozzle.

Darth Vader, the black-clad, breath-masked chief villain of the story and the Empire's most ruthless agent, flies a modified version of the standard TIE fighter. On his customized ship, the side solar panels are angled in at the top and bottom, reducing the size of the total configuration for better maneuverability in tight situations. An added aft section allows for twice the normal engine capacity. Vader's ship was designed this way by Industrial Light and Magic, not only to intensify his sinister aura but—for the sake of the audience—to render his ship distinguishable from other TIE's during the Death Star donnybrook. Vader is the only Empire personage to escape, tumbling off into outer space after his fighter is knocked out of control. (Without being told, the audience instinctively knows he'll be heard from again.)

An interesting sidelight on all this is the strange status of space-age technology presented in the story. Though there are deflector shields, land speeders, tractor beams, unbelievably sophisticated androids, holographic messages, hyperspace propulsion devices, and other evidences of advanced know-how, there are no guided missiles, even of the garden variety. In the frequent shootout confrontations, whether up close or over considerable distances, attackers and defenders act like gunslingers in Class B Westerns—forever aiming, shooting, and missing. It's as though cavemen had invented differential calculus but persisted in carrying stone handaxes. Perhaps this can be explained by the fact that the action takes place in the distant past rather than the distant future, and the levels of knowledge extant, though stratospherically high, are as spectacularly uneven as a roller coaster.

The designs of the X-wing, Y-wing, and standard TIE fighters (as well as those of the Imperial Star Destroyer and the rebel blockade runner) originated with Colin Cantwell. According to him, the X-wing's and Y-wing's atmospheric capabilities required a horizontal wing configuration for flight near planetary surfaces. Once into space, he had the former's wings "open up" for two reasons, visual excitement and easy identification.

In designing the evil-looking TIE fighter, he tried to mirror what he thought the mindset of a totally alien culture might be. Darth Vader's customized TIE fighter was not one of his designs, but spontaneously evolved during production as a convenient means of pinpointing the Vader ship when it zapped about on the screen.

Cantwell built prototype "study" models of his designs out of styrene and model-kit parts. Later these were translated into detailed drawings by Joe Johnston. The actual miniatures used for filming were painstakingly assembled under the direction of Grant McCune, chief model-maker of Industrial Light and Magic of Van Nuys, California, the firm that revolutionized special effects photography during the course of filming Star Wars (Special Photographic Special Effects Supervisor John Dykstra won an Academy Award, as did ILM's

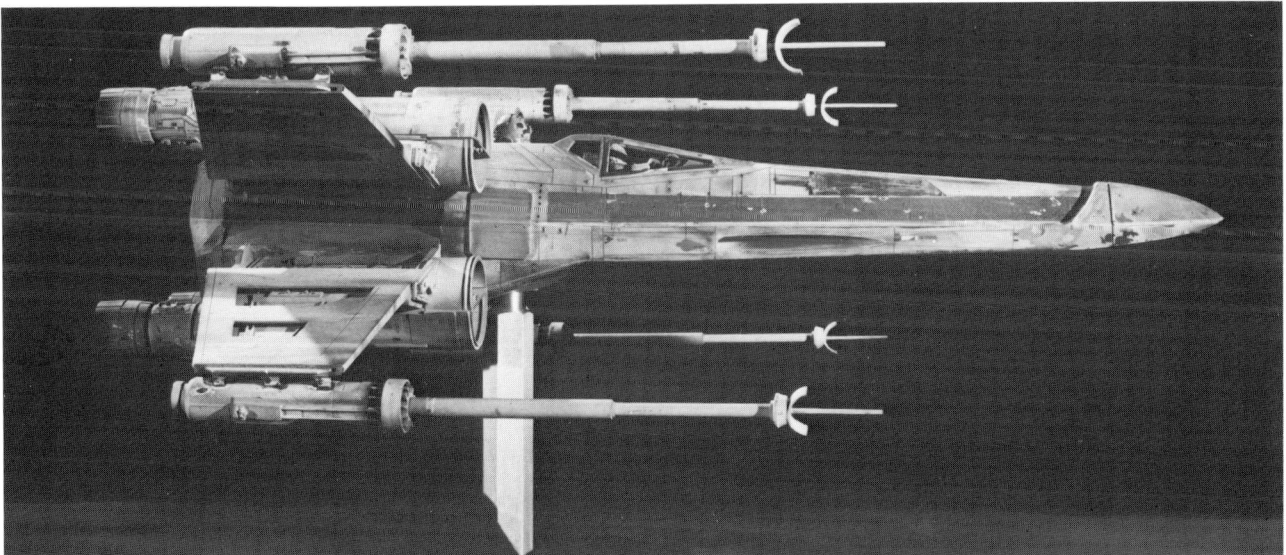

Richard Edlunds for photography, Robbie Black for opticals, and McCune for the miniatures).

From the Cantwell originals and Johnston drawings, McCune "edited" the actual film miniatures for the "stage." Such considerations as size, shape, and interior room for electronic devices influenced the construction of the miniatures. For example, as conceived, the X-wing was a sleeker, smaller ship. However, its aft section and wings had to be enlarged to accommodate the motors and wires that enabled the wings to flare open. Because of this, larger engines were added for symmetrical balance.

Of the five X-wing stage models, three had motorized wings. All five had air-cooled quartz lights in each of the four engines, plus LED (light-emitting diodes) cockpit interiors. To provide focal points for optics, which were later added to the film, lights were fitted into each wingtip laser cannon and fuselage-flanking torpedo tube.

The miniatures were built using parts from original molds and parts salvaged from hundreds of surplus plastic kits. Each model's surface was non-reflective, and of a special color, so as not to interfere with the blue-screen filming process. A Johnston sketch envisioned a bubble canopy for the X-wing, but a flat glassless canopy had to be used to avoid a light reflection that would have been detrimental (by showing a hole in the ship) to the blue-screen process.

Although the X-wings and Y-wings were weathered because Director George Lucas wanted them to look real, McCune also noted, "It is a lot easier to build a dirty model than a clean one."

Painting procedures were radically different from ones plastic modelers might use. The X-wings were first painted with a white enamel automotive primer to create a hard undersurface that couldn't be scratched away. Liquid frisket was dabbled about with a paintbrush, and then the models were sprayed with Floquil Railroad colors—sometimes with airbrushes that hadn't been cleaned for weeks. With removal of the frisket, the undersurface showed through to resemble areas of extreme wear. The red wing-stripes were added at the suggestion of Johnston to help tell one X-wing apart from another.

Finally, grease and grime colors were dipped into thinner and flowed across the model with a brush. Steel wool was used to create further weathering and rub down areas that needed to be redone. "Sometimes," admitted McCune, "you're hard put to know when you've gone far enough."

Nine X-wings were made from silicone rubber molds and back-cut for use in scenes in which they had to explode. Finalized model designs were sent to John Barry, at EMI Studios in London, England, to be used as guides in building full-scale ships for hangar scenes.

All the miniatures had built-in armatures that could be attached to an external motion device—at a variety of angles—to simulate in-flight movement. One modeler asked about the location of the entrance hatch on the TIE fighter. The studio models had three hatches—top, bottom, and back. That's because the models had mounting holes in those positions, and the hatches became a convenient way to cover the two out of three mounting holes not being used during filming.

The TIE fighters were painted a light gray-blue, almost a powder blue. However, because gray values are lost in film duplicating, the TIEs appeared much whiter on moviehouse screens.

Five stage versions were made of TIE fighters, including Darth Vader's. Each had LED-lit interiors and engine effects. The engine effects were added more as a device to give the fighters direction on the screen than as a preconceived part of the design. Sixteen TIEs of the exploding type were also built.

The final tribute to the model builders and special effects people might well be the fact that the nicknames ("studio jargon") that they coined for the ships for ease of reference spread like wildfire through the public awareness. Thanks to them we have such colorful appellations as X-wing and TIE—names that never were heard in the movie. One nickname, however, that did not escape beyond ILM walls was the jargon for the *Millennium Falcon:* the "pork burger."

INDEX

Apollo, 5, 13, 34
Assembly
    Bird of Prey, 47
    Command module, 34
    Lunar module, 34
    Raider, 86
    Saturn V, 28
    X-wing, 71

Base or stand
    Bird of Prey, 47, 50
    Enterprise, 57
Ballast, mounting, 57
Battle damage (Viper), 89
Battlestar Galactica, 80, 85
Bird of prey, 47
    assembly, 47
    base or stand, 47
    decaling, 50
    drybrushing, 49
    finishing touches, 50
    painting, 48
    using a wash, 49
Body modifications *(Galileo),* 63

Cockpit detailing, 77
Command module (CM), 6, 34
Cylon attack ship, 81

Decaling, 41, 50
Detailing, 72, 91
Drilling, 44
Drybrushing, 49

Eagle (lunar module), 13
Engine pods *(Galileo),* 66
Engine, hull, markings *(Galileo),* 65
Enterprise, 37, 39, 41, 58
    mounting ballast on, 57
    fiber optics on, 55
    lighting, 54, 68
    "Next Generation," 51
    painting, 56, 61
    production *Enterprise,* 38
    pilot film version, 38
    publicity *Enterprise,* 67
    saucer, 55, 56, 59
    secondary hull, warp engines, 60
    TV watcher's version, 41
    wiring, 54

Fiber optics, 45, 55, 85
Finishing touches
    Bird of Prey, 50
    Command module, 36
    Lunar module, 17, 36
Fluorescent tubes, used in
    Enterprise, 52

Galactica, 81
Galileo, 63
    instruments, 64
    landing gear, 66
    painting, 65
    windows, 64

Imperial Star Destroyer, 96
Imperial TIE (Twin Ion Engine)
    fighter, 76, 100
    cockpit detailing, 77
    masking, 76
    painting, 76
Instruments, recalibrating, 64

Junk, 91

Klingon Battle Cruiser, 43
    drilling, 44
    fiber optics, 45
    painting, 46
    wiring, 44
Klingon Bird of Prey. *See* Bird of Prey

Landing gear, 66
Lighting, 54, 68, 85
Lunar module (LM), 7, 13, 34
    ascent stage, 14
    descent stage, 16

Masking, 31, 76
Millennium Falcon, 96
Moon missions, 7
    chronology, 10
Mylar, 14, 36

Painting
    Bird of Prey, 48
    Command module, 35
    Enterprise, 56, 61
    Galileo, 65

junk, 91
Klingon Battle Cruiser, 46
Lunar module, 35
Saturn V, 29
TIE fighter, 76
transporter tug, 40

Planning (Klingon Battle Cruiser), 44

Raider, 80, 85
    fiber optics in, 85
    fuselage assembly, 86
    lighting, 86
Rebel blockade runner, 95

Saturn V, 19, 27
    assembly, 28
    launch preparations, 23
    masking, 31
    painting, 29
    statistics, 24
Saucer *(Enterprise),* 55, 56, 59
Secondary hull, warp engines *(Enterprise),* 60
Service module (SM), 6, 34
Soviet space program, 26
Star Trek, 37, 39, 41, 47, 58, 63, 67
Star Trek: The Next Generation, 43, 51
Star Trek II: The Wrath of Kahn, 59
Star Wars, 71, 76, 93

TIE fighter. *See* Imperial TIE (Twin Ion Engine) fighter
Transporter tug, 39
    painting, 40

Viper, 81, 88

Windows, 64
Wiring, 44, 54

X-Wing fighter, 71, 97
    assembly, 71
    detailing, 72
    weathering, 72

Y-Wing fighter, 98